Basic *Populus* Models of Ecology

To my father, who can figure out anything; from the cavity resonance of a magnetron, to building a house, to optimizing the acoustics of a violin.

Basic *Populus* Models of Ecology

Don Alstad
University of Minnesota

Prentice Hall

Prentice Hall, Upper Saddle River, NJ 07458

Library of Congress Cataloging-in-Publication Data

Alstad, Donald N.
   Basic Populus Models of Ecology / Donald N. Alstad.
      p. cm.
   ISBN 0-13-021289-X (pbk.)
      1. Ecology--Computer simulation. 2. Population biology--Computer simulation. 3.
   Populus. I. Title.

QH541.15.S5 A48 2001
577.8'80113--dc21

00-062372

Executive Editor: Teresa Ryu
Editorial Assistant: Colleen Lee
Production Editor/Page Layout: Kim Dellas
Manufacturing Manager: Trudy Pisciotti
Manufacturing Buyer: Michael Bell
Art Director: Jayne Conte
Cover Designer: Bruce Kenselaar

©2001 by Prentice-Hall, Inc.
Upper Saddle River, NJ 07458

Printed in the United States of America

10 9 8 7 6 5 4 3 2 1

ISBN 0-13-021289-X

Prentice-Hall International (UK) Limited, London
Prentice-Hall of Australia Pty. Limited, Sydney
Prentice-Hall Canada Inc., Toronto
Prentice-Hall Hispanoamericana, S.A., Mexico City
Prentice-Hall of India Private Limited, New Delhi
Prentice-Hall of Japan Inc., Tokyo
Pearson Education Asia Pte. Ltd.
Editora Prentice-Hall do Brasil, Ltda., Rio de Janeiro

Contents

Preface   vii

1   Density-Independent Population Growth   1

1.1 Exponential Growth with Continuous Breeding   2

1.2 The Geometric Growth of Discrete Cohorts   4

1.3 Comparing Continuous and Discrete Models   5

1.4 Simulating Density-Independent Population Growth with *Populus*   6

1.5 Postscript   10

References   12

Problems and Exercises   12

2   Density-Dependent Population Growth   16

2.1 The Continuous Logistic Model   16

2.2 Lagged Logistic Population Growth   25

2.3 Discrete Logistic Models   29

2.4 Detecting Density Dependence   33

References   35

Problems and Exercises   36

3   Demography and Age-Structured Population Growth 41

3.1 Estimating the Survival and Fertility Parameters
S_x, l_x, m_x, and F_x 42

3.2 Population Projection From the Life Table   46

3.3 Constant $l_x m_x$ Schedules   49

3.4 Age-Structured Population Growth   52

3.5 Cole's Paradox and the Evolution of Life Histories   55

3.6 Reproductive Value, V_x 57

3.7 Projection Matrices   37

3.8 Stage-Structured Populations   62

References   65

Problems and Exercises   66

4   Lotka-Volterra Competition   71

4.1 Empirical Examples   71

4.2 Lotka-Volterra Competition   73

4.3 Dynamics of Lotka-Volterra Competition   74

4.4 Isocline Analyses   77

4.5 Coexistence or Displacement?   79

4.6 Carrying Capacities and the Intensity of Competition
Affect the Probability of Coexistence   84

4.7 Postscript   86

References   86

Problems and Exercises   87

5   Continuous Predator-Prey Models   91

5.1 The Lotka-Volterra Predator-Prey Model   92
5.2 The θ-Logistic Model   104
References   111
Problems and Exercises   112

6   Infectious Microparasitic Diseases   115

6.1 *SI* Model with Density-Dependent Transmission   116
6.2 Dynamics and Equilibria   117
6.3 *SI* Model with Frequency-Dependent Transmission   122
6.4 *SI* Dynamics with Frequency-Dependent Transmission   122
6.5 *SIR* Model with Density-Dependent Transmission   125
6.6 Dynamics of Immunization   126
6.7 The Evolution of Virulence   129
References   130
Problems and Exercises   131

Glossary   135

Index   139

Preface

Population biology is a quantitative science dealing with changes in the size and composition of populations, and population biologists often use mathematical models to infer population dynamics. These models use information about the properties of individuals and basic assumptions about their interactions to predict population size, gene frequency, and optimal behavioral strategies of individuals, forming an important conceptual framework for courses in ecology, evolution, and behavior.

Students bring a range of quantitative backgrounds to the study of population biology. Although the models taught in introductory courses are not difficult, math anxieties and the abstract nature of the equations impede many students. The *Populus* software addresses this problem. Computer simulations provide a visual demonstration of model dynamics, helping students develop an intuitive connection between the form of the equations, their quantitative parameter values, and the resulting model behavior.

The first rudiments of *Populus* were produced in collaboration with my Minnesota colleagues Peter Abrams, Jim Curtsinger, and Dave Tilman, with 1986 funding from IBM, and integrated into a functional package by a superb student programmer, Chris Bratteli. We used the program at the University of Minnesota, shared it with a few friends elsewhere, and gradually improved it, supported by the National Science Foundation. In 1991, with 30 *Populus* models spanning introductory courses in ecology and evolution, I began to distribute the software and to encourage *gratis* duplication. Chris and I continued to develop *Populus* with a second programmer, Elizabeth Goehring, until 1995, when my attention turned to other interests.

The penultimate DOS version of *Populus* (3.4) became widely used, and pressure to revise it escalated as microprocessors with clock speeds beyond the capacity of Borland Turbo Pascal began to cause runtime errors. Together with a third programmer, Lars Roe, I began to rebuild *Populus* for Java virtual machines in summer 1998. We also patched version 3.4 to fix the Borland runtime error, producing version 3.42 for DOS windows in Win95, Win98, and WinNT. *Populus* 3.42 is available for download from *http://ecology.umn.edu/populus*. The Web site also provides full instructions.

The new Java *Populus*, version 5.x, is available from the same Web site, *http://ecology.umn.edu/populus*. This program runs as a local application (not an applet) on any platform that supports a Java Virtual Machine, including both Windows and Macintosh computers. Linked sets of input and output windows in the new interface allow toggled parameter increments to show up instantly on the appropriate graphs, so the effect of changing parameter values is immediate and clear. The current Java *Populus* version supports all of the basic ecological models described in this book and several basic evolutionary models. We will update this posting as the new package grows to full equivalence with DOS *Populus* 3.42 and beyond.

Minnesota ecology classes invest about a third of their lab meetings in model simulation. The remaining two thirds are devoted to experiments, foraging games, data manipulation, and role-playing discussion. Students first encounter a mathematical model in lecture after a series of empirical examples that motivate its central issues. I derive the equations and explain their significant properties before students come to the computer lab, then use simulation exercises to reinforce this initial exposure. We have found this sequence essential; students who attempt simulations without the conceptual preparation provided by a lecture or reading assignment take less away from the experience. This book should foster that conceptual preparation. It will supplement any of the major texts, guiding lab experience with the mathematical models of an introductory ecology course.

Sessions in the computer lab are structured around a series of questions designed to exercise pedagogically important features of the model. Problems and exercises in this book exemplify the conceptual focus that provides a good skeleton. We often begin and end the computer lab with a brief summary and questions, but students should have a great deal of freedom to experiment with parameter values and running conditions. *Populus* works best as a teaching tool when students are able to use it almost like a video game, testing "what if" scenarios.

Here are a few additional suggestions based on our Minnesota teaching experience with *Populus* simulations:

- It is helpful to have students work in pairs in a facility where they can discuss what they are seeing on the screen.
- Teaching assistants and faculty play an important role by looking over shoulders and engaging the students in discussion about the dynamics they are observing. The layout of the room must allow this circulation.
- When staff members see an instructive simulation, it is useful to have the whole class run that case and to give a brief explanation.
- Students need space to lay out their notebooks and papers around the computer.
- It is valuable to have the computer lab available for student use outside of the formal lab session and to make sure that each student with a computer has a personal copy of *Populus*.
- Resist the temptation to squeeze consideration of three or four mathematical models into one long lab session.
- Two dozen students, a dozen computers, one teaching assistant, one professor and a well-ventilated room make a good computer-lab combination. Brain function is inversely proportional to the number of hot computers and hot bodies crowded into a small space!

Simulation models in the *Populus* software cover subject matter from several course venues, including introductory ecology, introductory evolution, theoretical ecology, and population genetics. We decided early that proper documentation would require separate books for the ecological and evolutionary models. Reviewers subsequently convinced us to subdivide still further, producing a small, inexpensive treatment of the models used in elementary courses, followed by a more comprehensive book for the sophisticated clientele. Accordingly, we defer interesting ecological models of demographic stochasticity, resource competition, discrete predator–prey systems, diffusion, metapopulations, and macroparasites to that larger book.

Many people have played a role in bringing the *Populus* software to its present state. First among them is Chris Bratteli, who worked with me throughout his baccalaureate career at Minnesota; his skill with Pascal brought the first three generations of the program into being and provided a superb foundation for subsequent development. Liz Goehring carried on ably with the final DOS versions. Lars Roe has created the new interface of the Java *Populus* and is translating our proven Pascal algorithms for this new programming environment.

Special thanks to population biologists Peter Abrams, John Addicott, Sonia Altizer, Dave Andow, Janis Antonovics, Graham Bell, Doug Boucher, Andy Clark, Rob Colwell, Hugh Comins, James Crow, Jim Curtsinger, Andy Dobson, John Endler, Mike Gilpin, Lou Gross, Mike Hassell, Bob Holt, Yoh Iwasa, Mark Kirkpatrick, Rich Lenski, Bruce Levin, Simon Levin, Robert May, Martin Novak, Mike Rosenzweig, Jon Seger, Ruth Shaw, Frank Shaw, Frank Stewart, and Dave Tilman. Each of these colleagues has taken time to help us with the implementation of particular models.

This book owes its existence to two editors at Prentice Hall. The persistence and vision of Executive Life Sciences Editor Teresa Ryu persuaded me to write it at a time when many other interests competed for my attention. It was easier to write the book than to dissuade Teresa! Production Editor Kim Dellas has ably seen the manuscript into print. Working with both of them has been a pleasure.

It has been my privilege to have many superb teaching assistants from the Ecology, Evolution and Behavior Graduate Program at the University of Minnesota, and many of the exercises and problems in this book trace their lineage to those collaborations. It is an honor to acknowledge Sonia Altizer, Paul Cabe, Alison Chubb, Julie Etterson, Sarah Hotchkiss, Kevin Johnson, Sue Lewis, Elena Litchman, David Lytle, Mike Sorensen, Stuart Wagenius, Sarah Webb, and Bethany Woodworth. They are gifted teachers.

My first and best reviewer is always my wife, Karen Oberhauser. She is joined by Sonia M. Altizer, Emory University; Kemuel Badger, Ball State University; Joel S. Brown, University of Illinois at Chicago; F. Stephen Dobson, Auburn University; John Endler, University of California-Santa Barbara; Aaron M. Ellison, Mount Holyoke College; T. Luke George, Humboldt State University; Douglas E. Gill, University of Maryland; Richard Halliburton, Western Connecticut State University; David Hogg, University of Wisconsin; Robert D. Holt, University of Kansas; Brian Inouye, University of California-Davis; Stephen H. Jenkins, University of Nevada, Reno; Mark McKone, Carleton College; Mark McPeek, Dartmouth College; Manuel A. Morales, University of Maryland; William Morris, Duke University; Larry Rockwood, George Mason University; Wendy E. Sera, Baylor University; Daniel Simberloff, University of Tennessee; Donald M. Waller, University of Wisconsin-Madison; and Henry M. Wilbur, University of Virginia. I thank each of these reviewers for helping me to improve the focus and clarity of the manuscript.

Don Alstad
University of Minnesota

Basic *Populus* Models of Ecology

CHAPTER 1
Density-Independent Population Growth

Populations are a basic unit of ecological study. Inherent to the work of many ecologists is the question of how, why, and at what rate populations change in **size** or **density** (which is population size per unit area) over time. We will begin our study of this question, and of ecological models in general, with a very simple model of density-independent population growth. Because every female is a potential contributor of new progeny, large populations can grow much faster than small ones. When we say that a population-growth model is density independent, we imply that the rate of population growth *per capita*, or per individual, is independent of density. This means that individuals do not interfere with the reproduction, development, or survival of their neighbors, and hence that the population lives in a paradise of unlimited resources. Obviously, paradise exists only in the imagination, but any natural population that suffers a catastrophic reduction to low density may experience a period of effectively unlimited resources and density-independent growth. Density-independent models are also useful as a baseline for thinking about ecological dynamics; we can begin with a simple mathematical formulation and then use modular terms to add realistic complications. In this way, the more elaborate models become collections of component parts whose individual dynamic effects are easy to understand.

We usually define a population as a group of individuals of the same species within a limited area. The **Orca** whales of Puget Sound, bacteria in a cup of yogurt, and humans on Earth could be populations, but these examples demonstrate that our definitions are often arbitrary. Whales move up and down the Pacific Northwest coast, and those in the Sound may be only a temporary assortment. Bacteria in a cup of yogurt may comprise several different species, and to call humans on Earth one population proposes a fusion that would miss much of what is interesting about them. In fact, we usually specify population limits in a way that suits our convenience and the biological questions at hand.

Two sets of counteracting processes affect population size: **birth** and **immigration** cause it to increase, while **death** and **emigration** cause it to decrease. To simplify our model of density-independent population growth, we can make four assumptions.

1. Immigration and emigration balance, leaving birth and death as the only determinants of population size.
2. All individuals are identical with respect to their probabilities of dying or producing offspring.
3. Reproduction is asexual; populations consist entirely of **parthenogenetic** females, so we can ignore complications associated with mating. For example, when individual males can mate with multiple females, male abundance may not limit population growth.

4. Environmental resources are infinite, so the only factors affecting population size are the organisms' intrinsic birth and death rates. This is the paradise assumption.

These assumptions allow an extremely simple model of population growth. We could drop any or all assumptions and deal with the biological issues they hide, but that would make our analysis more complex. Instead, we will start with a truly basic model, presenting it in mathematical formats suitable for organisms with two different kinds of life histories.

1.1 CASE I. EXPONENTIAL GROWTH WITH CONTINUOUS BREEDING

First let's consider organisms like *Homo sapiens* or bacteria in a culture flask, with continuous breeding and overlapping generations. All ages will be present simultaneously, and population size will change steadily in small increments with the birth and death of individuals at any time. We model this **continuous** population growth with a **differential equation**, using **instantaneous rates** applied over infinitely small intervals of time. To model the size of a continuously changing population, we will define three model **parameters**.

$N(t)$ = population size, which is a continuous **function** of time, t

b = the instantaneous birth rate per individual

d = the instantaneous death rate per individual

We can represent the changes of population size that will occur under these assumptions and definitions with a differential equation:

$$\frac{dN}{dt} = (b - d)N \tag{1.1}$$

The expression on the left side, dN/dt, is the differential of N over the differential of t. Their ratio is the derivative of N with respect to t. It is an expression from calculus, meaning the rate of change of population size (N) with respect to time (t). Notice that the "d" signifying a differential is not printed in italics. The right side of equation (1.1) is a product of population size, N, multiplied by the difference between instantaneous birth and death rates. If the difference is positive, change in population size will be positive; if the difference is negative, change in population size will be negative. Here, the instantaneous birth and death rates, b and d, are italicized, as model parameters will be throughout the book. We can combine the instantaneous birth and death rates into a single parameter, $r = (b - d)$, called the **intrinsic rate of increase** or **instantaneous growth rate**. Our equation then becomes

$$\frac{dN}{dt} = rN \tag{1.2}$$

This expression states that the rate of change of population size with respect to time is proportional to N and the instantaneous growth rate, r. When $r = 0$, birth and death rates balance, individuals just manage to replace themselves, and population size, N, remains constant. When $r < 0$, the population shrinks toward extinction, and when $r > 0$, it grows larger. If we divide the whole population-growth rate, dN/dt, by population size, N, then we have the growth rate *per capita* (per individual),

$$\frac{dN}{Ndt} = r \tag{1.3}$$

so r is also called the ***per capita* growth rate**. As an **instantaneous rate**, r quantifies the *per capita* change in population size over an infinitely small time interval. To give r a numerical value, we must scale it over some finite, observable period, so the units will be individuals per day, per year, etc. We will revisit this issue later in the chapter.

Equations (1.1) through (1.3) specify the rate at which population size changes. Often what we want to know is not the rate of change, but population size itself as a function of time, so that we can project future population sizes. To do this, we add up successive changes by **integrating** the differential equation over time. Beginning with equation (1.2) the first steps are rearrangements:

$$dN = rN\, dt \tag{1.4}$$

$$\frac{dN}{N} = r\, dt \tag{1.5}$$

Then we **integrate**, summing changes in population size over the interval from t_0 to t,

$$\int_{t_0}^{t}\frac{dN}{N} = \int_{t_0}^{t} r\, dt \tag{1.6}$$

$$\ln N(t) - \ln N(t_0) = rt - rt_0 \tag{1.7}$$

where ln stands for the **natural logarithm**. Finally, we rearrange again to get a more interpretable expression,

$$e^{\ln N(t) - \ln N(t_0)} = e^{rt - rt_0} \tag{1.8}$$

where $e \cong 2.71$, the base of the natural logarithms. Continuing the rearrangements,

$$\frac{N(t)}{N(t_0)} = e^{rt - rt_0} \tag{1.9}$$

and if $t_0 = 0$, then

$$N(t) = N(t_0)e^{rt} \tag{1.10}$$

Equation (1.10) allows us to project future population size, $N(t)$, from a knowledge of the constant growth rate, r, the present population size, $N(t_0)$, and the number of intervals, t, over which growth occurs.

1.2 CASE II. THE GEOMETRIC GROWTH OF DISCRETE COHORTS

Now let's consider a density-independent growth model that is more appropriate for plants, insects, mammals, and other organisms that reproduce seasonally. Individuals in such populations comprise a series of **cohorts** whose members are the same age. Assume that a time interval begins with the appearance of newborns and that surviving individuals produce another cohort of offspring at the beginning of the next interval. Initially we will also assume that parents may survive to reproduce again so that generations are partially overlapping (like many mammals). The young appear in nearly synchronous groups separated by intervals without recruitment of new individuals into the population. For example, fawns of the white-tailed deer, *Odocoileus virginianus,* are born in the spring, so an annual fall census would reveal animals to be 0.5, 1.5, 2.5, 3.5 years old, and so on.

We can use a periodic census to quantify **discrete** changes in the population size of organisms with overlapping cohorts and model the changes with a **finite difference equation**. A difference equation projects population size over a finite, observable time interval from the birth of one cohort until the birth of the next. To illustrate, assume that individuals must survive to the end of a cohort interval to reproduce and that:

N_t = population size at time t;

b = births per individual during one time interval;

p = an individual's probability of surviving from one census to the next.

These parameters of the discrete model are all different from their analogs in the continuous example of Case I. Population size, N, is no longer a continuous function of time; the t is subscripted to indicate that N_t is the population size observed or expected in a census at time t. The b and p are finite rates, representing the expected number of births and probability of surviving per individual over an observable interval until the next census.

Using these parameters, we can project population size over the interval from time $t-1$ to time t as follows:

$$N_t = pN_{t-1} + pbN_{t-1} = (p + pb)N_{t-1} \tag{1.11}$$

We can also define a new parameter for the parenthetic term collecting the birth and survival rates, $\lambda = (p + pb)$, just as we did in Case I. This λ (a lower-case Greek lambda) is the number of survivors and progeny derived from each individual that appeared in the census at the beginning of the interval. For a strictly annual plant species, no adults survive significantly beyond the maturation of new seeds; in this case, with no overlap between generations, it would be more appropriate to define $\lambda = pb$. It is also easy to envision variations (like seed dormancy) that would require us to use different survival rates for different life-history stages. We will deal with these complications in a later chapter on demography and age-structured population growth.

Substituting λ to simplify equation (1.11), we can project population size backward (or forward) over one or more time intervals.

$$N_t = \lambda N_{t-1} = \lambda(\lambda N_{t-2}) = \lambda(\lambda(\lambda N_{t-3})) = \cdots = \lambda^t N_0 \tag{1.12}$$

We call λ the **discrete-** or **finite-growth rate**, but more properly, λ is a multiplicative growth factor that determines the proportional change in population size over a discrete interval. If $\lambda = 1$, then individuals just manage to replace themselves and population size at the next census remains constant. If $\lambda < 1$, the population shrinks toward extinction, and if $\lambda > 1$, it grows larger. As long as λ remains constant (an implication of the paradise assumption), we can predict future population sizes using the growth factor, λ, the present population size, N_0, and the number of intervals, t, over which growth occurs, with the equation

$$N_t = \lambda^t N_0 \tag{1.13}$$

The population grows in a series of discrete steps. Its size increases more rapidly as the number of organisms rises because all of the individuals already present can contribute new offspring.

1.3 COMPARING CONTINUOUS AND DISCRETE MODELS

If r and λ are both constants that determine the rate of population growth, how are they related? To illustrate, we can calculate the time required for a population to double with both models.

Case I. Continuous Growth

$$N(t) = e^{rt}N(t_0) = 2N(t_0) \tag{1.14}$$

$$e^{rt} = 2 \tag{1.15}$$

$$rt = \ln 2 \tag{1.16}$$

$$t = \frac{\ln 2}{r} \tag{1.17}$$

Case II. Discrete Growth

$$N_t = \lambda^t N_0 = 2N_0 \tag{1.18}$$

$$\lambda^t = 2 \tag{1.19}$$

$$t \ln \lambda = \ln 2 \tag{1.20}$$

$$t = \frac{\ln 2}{\ln \lambda} \tag{1.21}$$

Equations (1.17) and (1.21) give the doubling time for continuously and discretely growing populations, respectively. Setting these two equations equal and solving for λ in terms of r, or for r in terms of λ yields

$$\frac{\ln 2}{\ln \lambda} = \frac{\ln 2}{r} \tag{1.22}$$

$$r = \ln \lambda \tag{1.23}$$

$$\lambda = e^r \tag{1.24}$$

This comparison assumes that the instantaneous growth rate, r, is scaled to the same finite, observable time interval (days, years, t, etc.) used to define the discrete growth factor, λ.

1.4 SIMULATING DENSITY-INDEPENDENT POPULATION GROWTH WITH *POPULUS*

The *Populus* software implements density-independent population growth models on the Models menu. Both the continuous and discrete versions run from the same *Populus* input window, illustrated in Box 1.1, with input parameter values and option

---

BOX 1.1

The *Populus* input window for Density-Independent Population Growth, with parameter values and option settings that produce Figure 1.1a.

Parameter boxes with up and down toggles can be clicked higher or lower, and changes show up instantly on the linked output graph. Values can also be typed into the box.

The same buttons for viewing output, accessing help, saving, printing, and closing occur in every *Populus* input window.

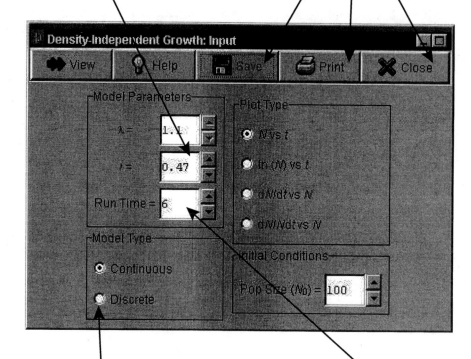

Continuous and discrete density-independent population growth models are both accessed from the same menu entry and input window.

Try increasing the run to 8, 10, 20 and 50 time steps. Watch the scale of the *y*-axis, and the increasing "j" shape of the growth curve.

settings that produce the continuous *Populus* output graph in (Figure 1.1a). Switching the Model Type option from continuous to discrete (Box 1.2) yields the corresponding output graph in Figure 1.1b. The initial population sizes in both graphs are $N(t_0) = N_0 = 100$, and the growth rates are $r = 0.47$ and $\lambda = 1.6$, respectively. The continuous version (Fig. 1.1a) models instantaneous changes in population size with a smooth exponential curve, and we call this exponential population growth. The discrete version (Fig. 1.1b) models population growth by projecting a series of periodic censuses, and we do not know what is happening between the successive estimates. For example, white-tailed deer populations may grow in the spring and shrink during the fall hunting season, but we would not see these details in a series of censuses made annually on New Year's Day. We call the resulting pattern of constant-multiple jumps geometric population growth. Any multiplicative process is *geometric*; *exponential* growth is a special case where changes are compounded instantaneously, and the time trajectory of population size is a smooth, continuous curve.

Equations (1.23) and (1.24) imply that continuous and discrete populations show the same change in size over a comparable period when $r = \ln \lambda$ or $\lambda = e^r$, and both rates are scaled to the same finite interval, t. *Populus* makes this rate conversion automatically with the switch between continuous and discrete models, using a discrete λ comparable to the instantaneous r of the previous run, or vice versa. For the continuous and discrete populations of Figure 1.1a and 1.1b, $\ln \lambda = \ln 1.6 = 0.47 = r$. Entering a new value in the appropriate rate-parameter window overrides this automatic conversion, initiating a new simulation independent of the previous setting.

Recall that if $r < 0$ or $\lambda < 1$, populations will decline toward extinction. The mathematical form of density-independent growth models dictates that this decline will approach $N = 0$ **asymptotically**. Consider the discrete case: if $\lambda < 1$, then λ^t will be closer and closer to zero as t increases. Since $N_t = \lambda^t N_0$, population size will never reach zero,

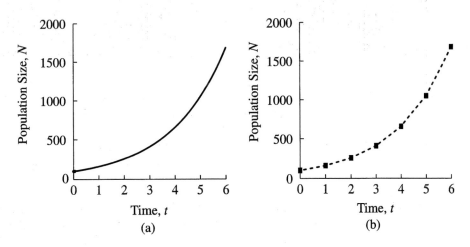

Figure 1.1   *Populus* simulations of continuous, exponential (a) and discrete, geometric (b) population growth with initial size, $N(t_0) = N_0 = 100$ and the growth rates $r = 0.47$ and $\lambda = 1.6$, respectively. The smooth curve in (a) gives a continuous projection of population changes with infinitesimal increments. In contrast, the dotted lines connecting discrete data points in (b) imply that we do not know what population size is between these censuses.

BOX 1.2

The *Populus* input window for Density-Independent Population Growth, with parameter values and option settings that produce Figure 1.1b.

> Switching from continuous to discrete growth yields a finite rate comparable to the previous instantaneous value.

> Switching from continuous to discrete growth also produces a new set plot types.

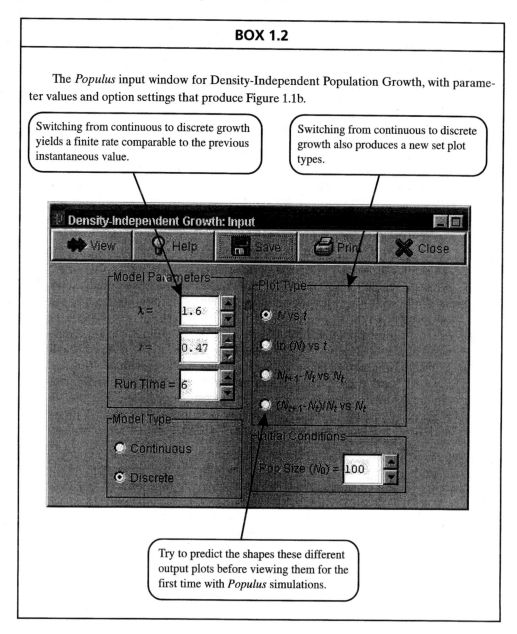

> Try to predict the shapes these different output plots before viewing them for the first time with *Populus* simulations.

---

but it does reach values of $N_t < 1$, which are effectively zero. We can illustrate the comparable continuous case beginning with equation (1.10). If r is negative, then the exponent rt is negative; i.e.,

$$N(t) = N(t_0)e^{-(rt)} = \frac{N(t_0)}{e^{rt}}$$

(1.25)

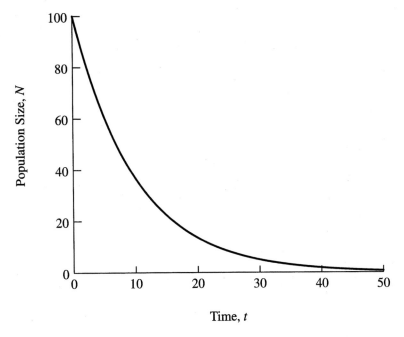

Figure 1.2   *Populus* plot of continuous exponential population decline from an initial size of $N_0 = 100$, with $r = -0.1$. Because population size falls by a constant multiple with each successive time step, changes in population size become very gradual as the population dwindles, giving an asymptotic approach to extinction.

As t increases and/or r grows more negative, the denominator on the right side of equation (1.25) grows larger; so again $N(t)$ approaches, but will never reach, zero. Figure 1.2 shows a *Populus* simulation of continuous exponential decline.

The Plot Type box in the Density-Independent Population Growth: Input window (Box 1.1) lists graphical options that illustrate several more aspects of the model. For example, it is instructive to take the natural logarithm of both sides in equation (1.10), yielding

$$\ln N(t) = \ln N(t_0) + rt \tag{1.26}$$

This is the expression for a straight line, so the semi-logarithmic plot of $\ln N(t)$ vs t (or a plot of N vs t with a logarithmic scale on the y-axis) is a straight line with slope r and a y-intercept of $\ln N(t_0)$. Figure 1.3 gives an increasing semi-log plot comparable to Figure 1.1a, a declining example comparable to Figure 1.2, and a static example for the case where $r = 0$. If we have empirical data on population growth, we can rearrange Equation (1.26) to estimate r from the difference between the logarithms of successive population sizes, $\ln N(t+1)$ and $\ln N(t)$,

$$r = \ln N(t_1) - \ln N(t_0). \tag{1.27}$$

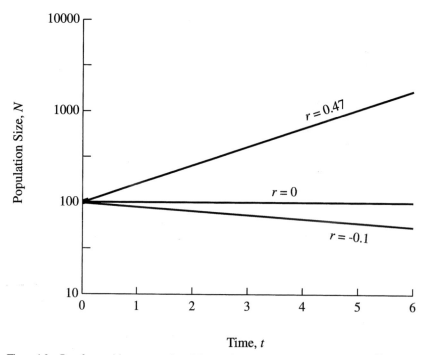

Time, t

Figure 1.3   *Populus* semi-log output plot of the continuous exponential case shown in Figure 1.1a with $r = 0.47$ and comparable cases with $r = 0$, and $r = -0.1$. Population size increases by a constant multiple with each succeeding time step. The y-axis of this semi-log plot is scaled so that each successively higher tic and value label are constant multiples of the previous one. In this case, the constant multiple is 10, but semi-log plots with a multiple of $e \cong 2.71$ are also common. In either case, the exponential curve of Figure 1.1a then becomes a straight line. Corresponding plots for static and declining population densities have flat and negative slopes. The declining case is exactly comparable to the example of Figure 1.2, which has a y-axis scaled in additive increments.

Additional plot options include graphs of both population growth, dN/dt or $N_{t+1} - N_t$, and *per capita* population growth, dN/Ndt or $(N_{t+1} - N_t)/N_t$, as function of N. Examples corresponding to the continuous simulation of Figure 1.1a appear in Figures 1.4 and 1.5. These graphs illustrate the important distinction between population growth and population growth per individual, and they provide an instructive counterpoint to their density-dependent analogs in the next chapter, Figures 2.6 and 2.2.

1.5 POSTSCRIPT

How well does a geometric or exponential model describe the growth of populations in the real world? The answer is that despite their simplicity, these density-independent models are not too bad at describing the growth of small populations. Problems arise when a density-independent model runs long enough to produce large population sizes. The time trajectory, N vs t, then becomes so strongly J-shaped that population size will quickly overflow the numerical capacity of our computers or the finite resources of our environment. In fact, with $\lambda > 1$ or $r > 0$, the density-independent growth expressions give mathematical descriptions of an explosion; a population growing in this manner

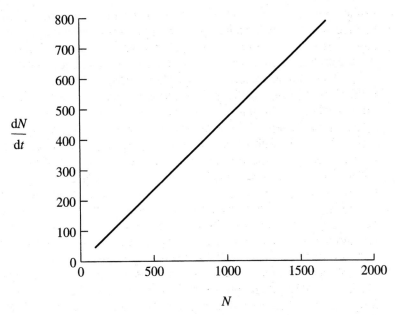

Figure 1.4 *Populus* plot of population growth, $\mathrm{d}N/\mathrm{d}t$, as a function of population size, N. Density-independent growth is directly proportional to N because every individual is a potential reproductive contributor and the population is assumed to live in a paradise of unlimited resources. Contrast this view with its density-dependent analog in Figure 2.6.

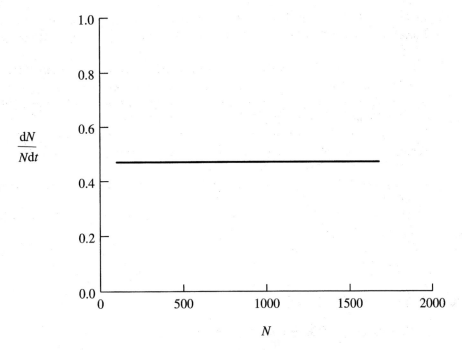

Figure 1.5 *Populus* plot of *per capita* population growth, $\mathrm{d}N/N\mathrm{d}t$, as a function of population size N. Contrast this view with its density-dependent analog in Figure 2.2.

would soon occupy the entire surface of the planet and squeeze itself to death (von Foerster *et al.*, 1960). The models offer a poor description of real population growth at high density because they incorporate an assumption that can never be true for long, that the environment is unlimited, so r (or λ) can remain constant.

What is the purpose of a population-growth model if it does not describe reality well? First, it illustrates the role of mathematics in population biology; math provides the logical rules that allow us to work out the implications and consequences of our assumptions. Second, the density-independent models provide a simple baseline. We can analyze their failures and then add modular components implementing environmental limitations, individual differences, sexual reproduction, or other real-life complications. We can then analyze the dynamic consequence of these components separately or in concert and their effect on the deficiencies of the baseline model.

Why have we produced two parallel and very similar analyses of density-independent population growth? First, because the differences reflect natural variation in real biological factors. In models incorporating **density-dependent feedback**, these life-history differences have striking effects on population-dynamic behavior, and comparison of discrete and continuous versions become quite interesting. Second, they illustrate two mathematical tools (continuous differential equations and discrete finite-difference equations) that are widely used in population biology.

REFERENCES

Case, T. J., *An Illustrated Guide to Theoretical Ecology*, New York: Oxford University Press, 2000, pp. 1–13.

Cohen, J. E., *How Many People Can the Earth Support?* New York: W. W. Norton & Co, 1995.

Ehrlich, P. R. and J. Roughgarden, *The Science of Ecology*, New York: MacMillan-Collier, 1987, pp. 65–75.

Elton, C., *The Ecology of Invasions by Animals and Plants*, London: Methuen, 1958.

Gotelli, N. J., *A Primer of Ecology*, Sunderland, MA: Sinauer Associates, 1995, pp. 1–23.

Krebs, C. J., *Ecology: The Experimental Analysis of Distribution and Abundance* (4th ed.), New York: HarperCollins College Publishers, 1994, pp. 198–204.

Ricklefs, R. E. and G. L. Miller, *Ecology*, 4th ed., New York: W. H. Freeman and Co., 1999, pp. 298–302.

Roughgarden, J., *Theory of Population Genetics and Evolutionary Ecology: An Introduction*, New York: MacMillan-Collier, 1979, pp 299–303.

Roughgarden, J., *Primer of Ecological Theory*, Upper Saddle River, N. J.: Prentice Hall, 1998, pp. 55–60.

von Foerster, H., P. M. Mora and L. W. Amiot, "Doomsday: Friday, 13 November, A.D. 2026," *Science* vol. 132, 1291–5.

Wilson, E. O. and W. H. Bossert, *A Primer of Population Biology*, Sunderland, MA: Sinauer Associates, 1971, pp. 92–102.

PROBLEMS AND EXERCISES

CONTINUOUS EXPONENTIAL GROWTH

1. If an exponentially growing population increases from $N = 5$ to $N = 25$ in one week, how long will it take to increase from 25,000 to 125,000?

2.   Look back at equation (1.10), and predict how a plot of N vs t would look when $r = 0.1, r = 0, r = -0.1$. Use *Populus* to simulate the growth of a continuous, exponential population with $r = 0.1$, an initial population size of $N(0) = 100$, and a run-time set to 50 time steps. Turn on the comparison-plotting function so that you can observe differences between this run and the next one, and then run a new simulation with $r = -0.1$. Finally, change to $r = 0$. How does the numerical value of r affect the change in population size over time?

3.   What would a plot of dN/dt vs N look like for an exponentially growing population? Use *Populus* to check your answer. Similarly, predict the shape and check a graph of dN/Ndt vs N. Explain in your own words how these graphs embody the assumptions on which the density-independent growth models are based.

4.   To get a feel for the tremendous growth capacity of a density-independent model, set r to 1.5, initial population size to 2, and the run-time to 20 time steps. Note the scaling of the y-axis on the output graph. Is the final population number larger or smaller than the current national debt?

5.   We can estimate r from consecutive population sizes. On the *Populus* parameter window for the density-independent growth models, set r to 0.5, initial population size to 20, and the run-time to 3 generations. Record the values of N from the output graph:

   $N(0) = 20$ $N(1) =$ $N(2) =$ $N(3) =$

   Estimate r by computing:

$$\ln N(1) - \ln N(0) = \underline{\hspace{2cm}}$$

$$\ln N(2) - \ln N(1) = \underline{\hspace{2cm}}$$

$$\ln N(3) - \ln N(2) = \underline{\hspace{2cm}}$$

   Use a pencil or calculator to estimate what population size will be at $t = 6$; then use *Populus* to check your answer.

DISCRETE GEOMETRIC GROWTH

6.   Look back at equation (1.13) and predict how a plot of N vs t would look when $\lambda = 1.1$, $\lambda = 1$, and $\lambda = 0.9$. Use the *Populus* program to simulate discrete, density-independent growth of a population with $\lambda = 1.1$, setting the run-time for 50 time steps and the initial population size at $N_0 = 100$. Turn on the comparison-plotting function so that you can observe differences between this run and the next one, and then run a new simulation with $\lambda = 0.9$. Finally, change λ to 1.0. How does the numerical value of λ affect the change in population size over time?

7.   What would a plot of $(N_{t+1} - N_t)$ vs N_t look like for a geometrically growing population? Use *Populus* to check your answer. Similarly, predict the shape and check a

graph of $(N_{t+1} - N_t)/N_t$ vs N_t. Explain in your own words how these graphs embody the assumptions on which the density-independent growth models are based.

8. We can estimate λ from consecutive population sizes. On the *Populus* parameter window for the density-independent growth models, set λ to 1.5, initial population size to 20, and the run-time to 3 generations. Record the values of N from the output graph:

$N_0 = 20$ $\qquad$ $N_1 =$ $\qquad$ $N_2 =$ $\qquad$ $N_3 =$

Then, estimate λ by computing:

$$N_1/N_0 =$$

$$N_2/N_1 =$$

$$N_3/N_2 =$$

Use a pencil or calculator to estimate what population size will be at generation 6; then use *Populus* to check your answer.

9. Compare the growth of two populations, one growing continuously from $N(t_0) = 100$ with an instantaneous growth rate of $r = 0.5$, and the second growing discretely from $N_t = 100$, with a finite growth factor of $\lambda = 2$. Which population will grow more rapidly? How long will it be before the faster-growing population is more than twice the size of the other? What assumptions did you make to solve this problem? Check your answers with *Populus* simulations.

PROBLEMS WITH EMPIRICAL DATA

10. California condors (*Gymnogyps californicanus*) are on the brink of extinction. By the early 1980s their numbers had dwindled to 21, and λ was estimated to be about 0.95 per year. At that point, all the remaining birds were captured for breeding programs at the San Diego Wild Animal Park, the Los Angeles Zoo, and the World Center for Birds of Prey in Idaho. By 1992 the captive population had grown sufficiently to allow releases back into the wild, and in January 2000 there were 53 condors living in California and Arizona and 105 in captivity or nearing release (*http://www.peregrinefund.org/notes_condor.html*). If the birds had been left in the wild with $N = 21$ and $\lambda = 0.95$, how long would it have been before the population declined to a single remaining pair? What assumptions did you make in answering this question? Use a density-independent growth model to project the geometric decline. Do you think that this is a good method of predicting the future of a small population? Why? What biological features and processes would you incorporate in a better model to help conservationists with the management of threatened populations?

11.   The *United Nations Demographic Yearbook* estimates that populations of *Homo sapiens* increased worldwide from 4.49×10^9 to 5.29×10^9 in the decade from 1980 to 1990. What was the continuous growth rate, r, per decade? Assume (a) that growth continues at the same density-independent rate after 1990 and (b) that there are about 1.49×10^{14} square meters of dry land (including the Edens of Antarctica, central Greenland, the Sahara, etc.) on the planet. How long will it be before the number of persons exceeds the number of square meters available for them to stand on? Guess before making the calculations with pencil and calculator, then run a *Populus* simulation to check your answer.

12.   Optimists will counter that growth rates cannot remain constant and hence that human population dynamics will depart from the explosive trajectory of a density-independent model. It is true that density-independent growth models do not fit the recent human data well, but the nature of the deviations may surprise you. Table 1.1 shows worldwide population estimates since 1950:

Table 1.1

UNITED NATIONS ESTIMATES OF THE HUMAN POPULATION WORLD WIDE (IN BILLIONS) FROM 1950 TO 1998.

Year	$N \times 10^9$	Year	$N \times 10^9$	Year	$N \times 10^9$
1950	2.52	1983	4.69	1992	5.48
1955	2.69	1984	4.76	1993	5.54
1960	3.02	1985	4.85	1994	5.63
1965	3.34	1986	4.92	1995	5.72
1970	3.70	1987	5.02	1996	5.77
1975	4.08	1988	5.11	1997	5.85
1980	4.49	1989	5.20	1998	5.90
1981	4.51	1990	5.29		
1982	4.59	1991	5.39		

Sources: 1950-1997, The United Nations Demographic Yearbook; 1998, *http://www.undp.org/popin/wdtrends.*

Use the data from 1950 and 1955 to estimate an r per five-year interval. Then set up a *Populus* simulation of continuous density-independent population growth with an initial population size of $N(t_0) = 2.52 \times 10^9$ and use this r to project population size 45 years later. How close is this simulation projection to the 1995 United Nations estimate? Make a paper print of the *Populus* output and pencil in the empirical data for comparison. Discuss the results with your classmates. Are there any other analyses that you would like to perform on these data? Students who would like to follow up on these issues should look at the short article by von Foerster, Mora and Amiot (1960) and Cohen's 1995 book, both listed with the references. The United Nations website *http://www.un.org* is another source for information and analyses of human demography.

CHAPTER 2

Density-Dependent Population Growth

The exponential and geometric models of population growth in Chapter 1 describe an explosive trajectory that is uncommon in natural populations. These simple models are density independent because they assume that populations grow at a constant rate under ideal conditions, in a paradise of unlimited resources. In fact, any population living in the real world is likely to grow more slowly as population size increases and resource supplies dwindle. Individuals may prosper with abundant resources when population size is small, but birth rates are likely to fall and death rates to rise as competition increases at higher density.

We now treat simple models of **density-dependent** population growth. The words density-dependent mean that *per capita* birth and death rates may change with population size. The usual mechanism (and principal focus of this chapter) is density-dependent competition; we incorporate a **negative feedback** in our models, causing *per capita* population growth to slow as population size (N) increases. Figure 2.1 gives an example of this negative feedback; the ovulation rates of white-tailed deer vary inversely with population size and their browsing impact on vegetation (Nixon 1965). It is also possible for feedback to be positive, i.e., for birth rates to rise and death rates to fall with increasing density. This positive feedback or facilitation may be common at low density. For example, prairie dog colonies may be warned of a predator incursion sooner when more watchful eyes are present. It may also be easier to find a mate at higher densities. Nevertheless, with the notable exception of *Homo sapiens*, most density-dependent feedback in moderate- or high-density populations is negative (Tanner 1966).

2.1 THE CONTINUOUS LOGISTIC MODEL

The continuous logistic model of density-dependent population growth shares several simplifying assumptions with the exponential model of Chapter 1. Specifically, we assume that (a) immigration and emigration balance, leaving birth and death as the only rate parameters governing population dynamics; (b) all individuals are identical; and (c) the population is asexual, so we can ignore complications associated with mating. For this density-dependent model, we drop the assumption that environmental resources are infinite, specifically implementing resource limitation and intraspecific (within-species) competition. Finally, we make one new assumption, that density-dependent feedback exerts its effect on population growth instantaneously.

We could add density-dependent feedback to a growth model in several ways. For example, we could redefine r to be a changing **function** of population size; but then r

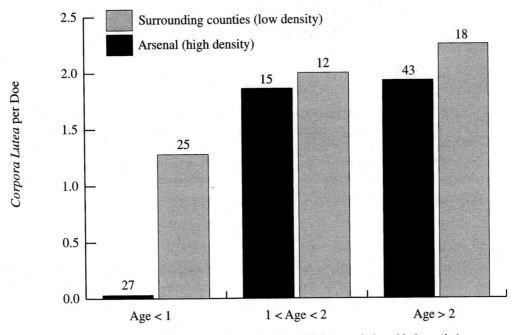

Figure 2.1   The number of *corpora lutea* (ruptured ovarian follicles, correlating with the ovulation, conception, and birth rates) carried by female white-tailed deer from low- and high-density populations in eastern Ohio. The high-density population within a fenced arsenal property had expanded to over-browse the vegetation. Hunting pressure in the surrounding area held the deer population at lower density. Numbers above each bar give sample sizes. This measure of ovulation was lower in the high-density population for each age class, with confidence > 95% (Nixon 1965). Reduced food availability in the over-browsed arsenal property also delayed the average age of first reproduction.

would lose its utility as a measure of the population's maximum intrinsic growth capacity. A more useful approach is to link population growth to resources and the competitive process that causes negative feedback. We can do this by defining K as the **carrying capacity** of the population's habitat, quantified in units of individual population members. We then need a mathematical formulation that will cause population growth, dN/dt, to decline as population size, N, approaches the carrying capacity, K. When $N = K$, population size has reached the habitat carrying capacity, and resource limitation should prevent further population growth. Conversely, a population with $N < K$ has not exhausted its supplies and should continue to grow. In this case, $K - N$ gives the unused capacity of the habitat, and $(K - N)/K$, is the unused fraction of carrying capacity, ranging from 0 (when $N = K$) to 1 (when $N = 0$).

We can use the fraction of carrying capacity remaining, $(K - N)/K$, as a feedback term to incorporate negative density dependence in a simple population-growth model. Let's start with the density-independent exponential model of Chapter 1.

$$\frac{dN}{dt} = rN$$

(2.1)

We can add negative feedback to this model by multiplying the right side by the fraction of unused carrying capacity.

$$\frac{dN}{dt} = rN\left(\frac{K - N}{K}\right)$$

(2.2)

This logistic model causes population growth, dN/dt, to change with population size, N. When $N = 0$, the feedback term has a value of 1, and the logistic equation, (2.2), is equivalent to the exponential model of equation (2.1). In contrast, when $N = K$, the feedback term has a value of zero, and population growth stops. To see how the logistic population-growth model behaves between the extremes of $N = K$ and $N = 0$, we can rearrange equation (2.2) as

$$\frac{dN}{Ndt} = r - \frac{r}{K}N$$

(2.3)

This is the equation for a straight line in the form $y = mx + b$; it shows that the *per capita* population growth rate, dN/Ndt, is a linearly declining function of N (Figure 2.2a). The *y*-intercept is r, the *x*-intercept is K, and $-r/K$ is the slope. With N near 0, population growth *per capita* is near the maximum intrinsic rate, r. As N increases, *per capita* growth falls steadily, reaching zero at $N = K$. Thus, population size has a negative feedback on population growth.

While a linear effect of N on dN/Ndt is the simplest possible negative feedback function, it is certainly not the only possibility. If individuals hoard resources so that

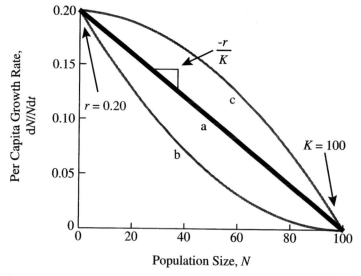

Figure 2.2 Growth rate *per capita*, declining as a linear function of N (line 3.2a) for a case with $r = 0.2$ (the *y*-intercept), and $K = 100$ (the *x*-intercept). The slope of this linear feedback function is $-r/K$. Curves show hypothetical feedback functions for interference competition (b), or territories of fixed size (c).

they interfere with their neighbors beyond the extent dictated by basic needs, then their competitive effects would be stronger at low density, and the feedback function might look like Figure 2.2b. Alternatively, territorial animals might have a feedback function like Figure 2.2c if competitive effects are weak until population density rises to the point where all potential territories fill. We will address the mathematical form of these nonlinear alternatives as a problem at the end of the chapter. Note, however, that they represent special cases. The linear feedback of a logistic model is a reasonable approximation when competition involves simple resource use. It also captures most of the properties of a more generalized negative-feedback formulation.

PER CAPITA VERSUS POPULATION GROWTH RATES

Notice that equation (2.3) and the y-axis of Figure 2.2 give the *per capita* population-growth rate, dN/Ndt. It is instructive to rearrange this equation once again, multiplying through by N to give the growth rate of the whole population,

$$\frac{dN}{dt} = rN - \frac{rN^2}{K} \qquad (2.4)$$

In this form, the equation shows that population growth, dN/dt, is a **quadratic** function of N. The roots, where $dN/dt = 0$, are at $N = 0$, and $N = K$. The graph of equation (2.4), showing dN/dt as a function of N, is parabolic (Figure 2.3). Population size is constant at the x-intercepts, $N = 0$ and $N = K$, for two different reasons. Near $N = 0$, *per capita* growth of the few individuals present is very high because $(K - N)/K$ is near its maximum (1), and there is very little density-dependent feedback (Figure 2.2). Nevertheless, the population as a whole grows slowly because very few individuals are reproducing. In contrast, when $N = K$, the population is large, but the *per capita* growth rate is 0 because $(K - N)/K$ is at its minimum (0), and density-dependent feedback is very strong; so again, the population-growth rate, dN/dt, is 0. It is important to distinguish between the *per capita* and whole-population growth rates graphed in Figures 2.2 and 2.3 and to compare the density-independent analogs in Figures 1.5 and 1.4.

The effect of density-dependent feedback on logistic population growth has practical implications. To manage a fishery or any other resource population so that it will replace our catch quickly and produce the maximum sustainable harvest, we should theoretically adjust consumption to hold N at $K/2$, the maximum of Figure 2.3. In practice, this theory is too simplistic because growth and recruitment of many populations, especially fish, vary with the age and size of members. The assumption that all individuals are identical ignores this essential biology. In fact, by holding N at $K/2$ we would reduce the average size and reproductive capacity of the fish and overexploit the population, reducing sustainable yield.

SIMULATING CONTINUOUS LOGISTIC POPULATION GROWTH WITH *POPULUS*

Populus implements Density-Dependent Population Growth on the Models menu. Three different versions (Continuous Logistic Growth, Lagged Logistic Growth, and

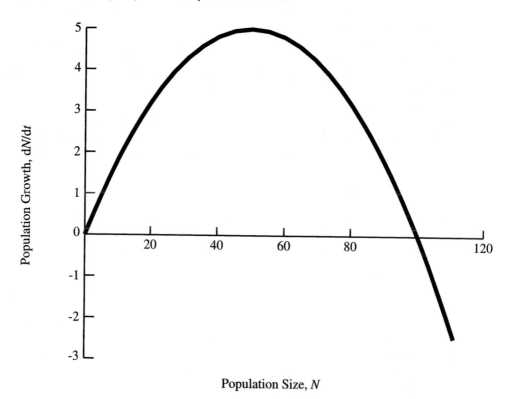

Figure 2.3   *Populus* output of population growth, dN/dt, as a function of N, for the same logistic population growth simulation illustrated in Figure 2.2a and Figure 2.4a ($K = 100, r = 0.2$).

Discrete Logistic Growth) all run from the same input window, illustrated in Box 2.1 with settings that produce the continuous logistic time trajectory (N vs t) of Figure 2.4a. With these same parameter values, clicking the button beside dN/N dt vs N in the Plot Type box produces the linear decline in *per capita* growth with population size shown in Figure 2.2a. The dN/dt vs N plot option produces the parabolic output graph in Figure 2.3, with a maximum sustainable yield at $K/2$.

DYNAMICS AND EQUILIBRIA

For a time trajectory of logistic population growth, we need to integrate the differential equation just as we integrated the simpler exponential model to produce the projection equation (1.10) in Chapter 1. The integration of equation (2.2) to give a solution for $N(t)$ is complex; students who would like to follow the details will find three different methods given by Roughgarden (1979), Emlen (1984), and Neuhauser (2000). All three methods yield the same result, that

$$N(t) = \frac{K}{1 + \left(\dfrac{K - N(0)}{N(0)}\right)e^{-(rt)}} \qquad (2.5)$$

BOX 2.1

The *Populus* input window for density-dependent population growth, with parameter values and option settings that produce the simulation of continuous logistic growth in Figure 2.2a.

Three different models of density-dependent population growth (the continuous logistic model, the lagged continuous logistic model, and the discrete logistic model) all run from the same input window.

Use the toggle button on the parameter box for the intrinsic rate of increase, r, to explore the interaction between growth rate and the time required to regain equilibrial density after a perturbation.

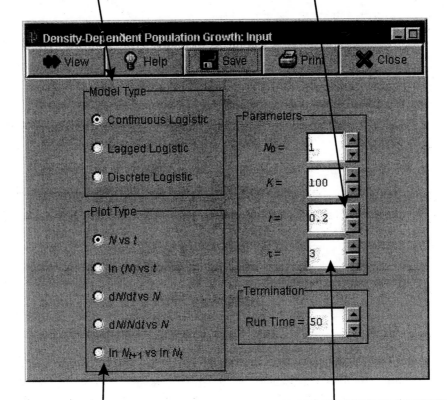

Each of these plot types makes a different conceptual point about density-dependent population growth. You should be able to predict their shapes and explain their differences.

The τ box is turned off and inaccessible for the continuous and discrete logistic models; it applies only to the lagged version.

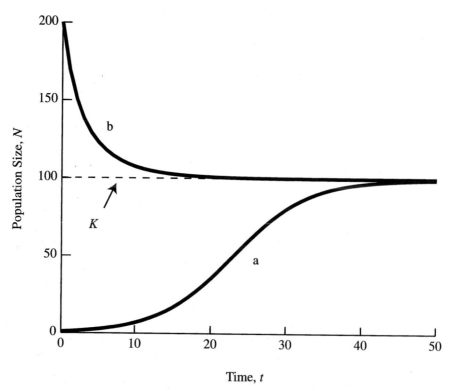

Figure 2.4   *Populus* output showing time trajectories, $N(t)$ vs t, for continuous logistic population growth. The lower, sigmoid curve (a) begins at $N(0) = 1$, with $r = 0.2$, and $K = 100$. The upper curve (b) starts above the carrying capacity, with $N(0) = 200$. Again, $r = 0.2$ and the carrying capacity, which is marked by a dotted line, is $K = 100$.

This equation is often rearranged, multiplying the right side by $N(0)/N(0)$ to eliminate the compound fraction,

$$N(t) = \frac{N(0)\,K}{N(0) + [K - N(0)]e^{-(rt)}} \tag{2.6}$$

A plot of $N(t)$ with respect to time gives the **sigmoid** (S-shaped) form of Figure 2.4a, where growth is nearly exponential when N is near zero and slows to equilibrium at $N = K$. When initial population size exceeds the carrying capacity as in Figure 2.4b, numbers fall in an **asymptotic** approach toward K. The sigmoid time trajectory of increasing population size (N vs t) reflects two opposing tendencies. The number of reproductive individuals in the population increases with N, but the reproductive output per individual decreases with N, limited by density-dependent feedback. Therefore, the maximal whole-population growth rate, dN/dt, occurs when these opposing processes are both at intermediate levels. $N = K/2$ gives the steepest slope and the inflection point of Figure 2.4a, and the maximum population growth rate in Figure 2.3.

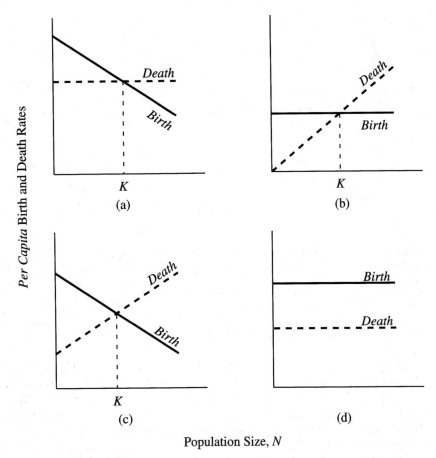

Figure 2.5 *Per capita* birth (solid line) and death (dashed line) rates as a function of *N*. Regulated equilibria require a density-dependent slope for one or both rates. In (a), the birth rate is density dependent, while the death rate is density independent. There is a regulated equilibrium at *N* = *K*, where the rates are equal. Above *K*, *d* > *b*, and population size declines. In (b), the death rate is density dependent, while the birth rate is density independent. Again, there is an equilibrium where the rates balance. In (c), both rates are density dependent. In (d), neither rate varies with density; since the birth rate exceeds the death rate, this case leads to unregulated exponential growth.

Because continuous logistic population growth tends toward *K* from any population size, *K* is a **globally stable equilibrium**. Such a population is **regulated** by the density-dependent feedback of population size on population growth. A regulated population that is perturbed, or shifted away from the equilibrium density, will return to *K*. If population density is reduced to a value less than *K*, either birth rates increase (Figure 2.5a) or death rates decrease (Figure 2.5b) or both (Figure 2.5c). If population density rises to a value above *K*, either birth rates decrease (Figure 2.5a) or death rates increase (Figure 2.5b) or both (Figure 2.5c). At *N* = *K*, birth and death rates are equal. Figure 2.5d shows a comparable example of birth and death rates for an exponentially

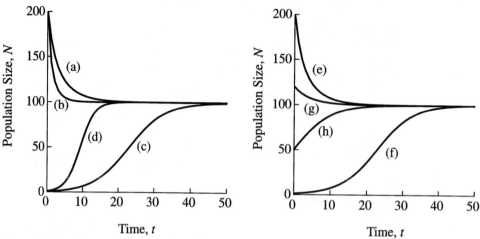

Figure 2.6 *Populus* output showing time trajectories, N vs t, for continuous logistic population growth. All eight simulations in this figure have the same carrying capacity, $K = 100$. Four cases on the left compare the return of populations with different growth rates to equilibrium. In runs (a) and (c) with $r = 0.2$, the perturbation to $N(0) = 1$ or $N(0) = 200$ takes longer to decay than runs (b) and (d), with $r = 0.5$. The graph on right illustrates the effect of starting at different distances from the equilibrium point. Here in all four cases $r = 0.2$. In cases (e) and (f), with perturbations to $N_0 = 1$ or $N(0) = 200$, the return to equilibrium takes longer than in cases (g) and (h), with smaller perturbations to $N(0) = 50$ or $N(0) = 120$.

growing population; because the rates do not vary with density, there is neither density-dependent feedback nor an equilibrium where birth and death rates balance.

The intrinsic growth rate, r, determines the speed with which a regulated population approaches equilibrium. When two populations receive the same perturbation away from their equilibrium density, the population with the higher r returns to K more quickly (Figure 2.6a). This process can be viewed on screen by toggling higher or lower r values in the input window (Box 2.1). Similarly, when populations with the same r receive different perturbations, the one that begins closer to K will return to equilibrium density sooner (Figure 2.6b). Again, this can be visualized by toggling higher or lower values of $N(0)$. To allow comparisons independent of starting point, it is helpful to define a characteristic return time, T_R, as the reciprocal of the intrinsic growth rate,

$$T_R = \frac{1}{r} \tag{2.7}$$

the characteristic return time suffices for a population with logistic growth to recover about 66% of a perturbation from the equilibrium density (Pimm 1991).

POPULATION DOUBLING TIME

We can parallel our calculation of the doubling time for exponential population growth in Chapter 2 by setting $N(t) = 2N(0)$ in equation (2.5) and solving for t (an estimate of doubling time for a population with continuous logistic growth).

$$2N(0) = \frac{K}{1 + \left(\dfrac{K - N(0)}{N(0)}\right)e^{-(rt)}} \tag{2.8}$$

$$2N(0) + [2K - 2N(0)]e^{-(rt)} = K \tag{2.9}$$

$$e^{-(rt)} = \frac{K - 2N(0)}{2K - 2N(0)} \tag{2.10}$$

$$t = \frac{1}{r}\ln\left[\frac{2K - 2N(0)}{K - 2N(0)}\right] \tag{2.11}$$

Equation (2.11) is more complex than the comparable expression we developed for an exponential model (equation (1.17)), where r was the only parameter. It should be no surprise that r, $N(0)$, and K are all necessary to predict the doubling time of a population with negative, density-dependent feedback. In fact, the logistic model only allows population size to double if $N(0) < K/2$ When $N(0) > K/2$ the denominator and hence the entire bracketed ratio of equation (2.11) becomes negative. Since the natural log of a negative number is undefined, the population will never double.

Two other features of the continuous logistic model deserve mention. One is the range of values that r can assume. With an exponential growth model, we could use both positive and negative values of r. The logistic population growth given by equation (2.2) is the product of r, N, and the feedback term $(K-N)/K$. If $N > K$, the feedback term is negative. If r is also negative because the death rate exceeds the birth rate, the product of those three terms is positive, and the equation gives an erroneous result; population growth increases indefinitely with N. To avoid this anomaly, we constrain use of the logistic model to cases where $r \geq 0$. Finally, with the continuous logistic model we assume that the density-dependent negative feedback of population size on population growth occurs instantaneously. This assumption has important implications, and we will now explore variants of the logistic model that focus on the timing of feedback.

2.2 LAGGED LOGISTIC POPULATION GROWTH

Changes in population density may not have an immediate effect on birth and death rates. For example, declining nutritional status caused by high consumer densities and decreased food availability may lower birth rates, but a generation may pass before changes in birth and juvenile death affect adult population size. Alternatively, many species depend on resources whose own density may take some time to rebuild after suffering heavy exploitation, even if the consumers decline. Either process would delay the regulatory effect of density-dependent feedback, causing it to operate with a time lag. In this section, we will analyze the effect of lagged, or delayed, feedback on the dynamics of logistic population growth. Figure 2.7 gives the *Populus N* vs *t* trajectory of a lagged logistic model first suggested by Hutchinson (1948, 1978) as follows:

$$\frac{dN}{dt} = rN(t)\left(1 - \frac{N(t - \tau)}{K}\right) \tag{2.12}$$

Hutchinson's equation is identical to the continuous logistic model of equation (2.2), except that now the feedback of population size, *N*, on population growth, d*N*/d*t*, depends on the population size at time *t* − τ. Here τ is a finite interval, so *t* − τ is some time prior to *t*. For many insect species, the amount of food consumed as juveniles (larvae) influences the subsequent reproductive output of adult females. If τ is the amount of time it takes to grow from larva to reproductive adulthood, then population growth rate at time *t* could be a function of population density at time *t* − τ. The lagged logistic option in the *Populus* input window for density-dependent population growth (Box 2.2) allows the same plot options available with the continuous logistic model. It also activates a parameter box for τ, which can be set from 0 to 5 time steps.

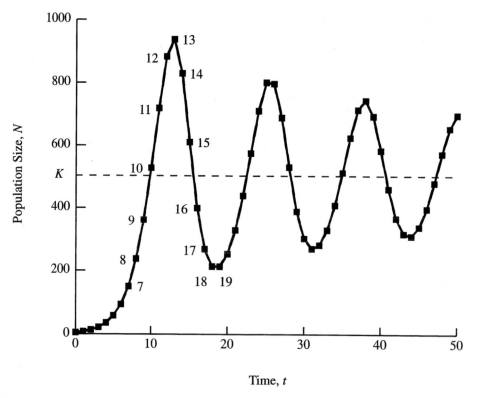

Figure 2.7 *Populus* output giving the population-size trajectory (*N* vs *t*) for continuous lagged logistic population growth with *K* = 500, and a lag of τ = 3 time steps. Initial population size for this run was *N*(*t*)= 5, and the intrinsic growth rate was *r* = 0.5. Numbers printed beside the plot give the time steps from the *x*-axis to illustrate the delayed feedback.

BOX 2.2

The *Populus* input window for density-dependent population growth, with parameter values and option settings that produce the simulation of lagged logistic growth in Figure 2.7.

> Choosing the lagged-logistic model option activates the τ window in parameter box.

> Use toggle buttons to experiment with different values of the rτ product.

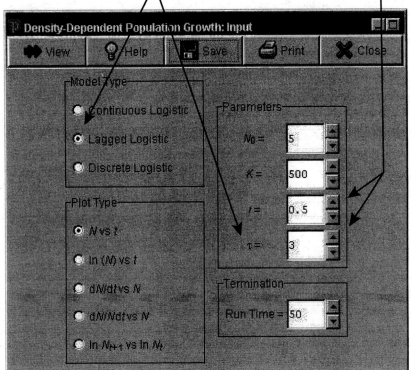

> The dynamics of this model often oscillate. Which parameters have the most influence over the period and amplitude of these oscillations?

DYNAMICS AND EQUILIBRIA

The addition of a time lag makes equation (2.12) too complex for a definite integral solution like equation (2.5). Instead, we use a computer to project the differential equation and tabulate successive changes in population size, a technique called numerical integration. To see how this model behaves, look at the numbered time steps in the *Populus N* vs *t* trajectory of Figure 2.7. At point $t = 10$, the growing population is just above K. If feedback were instantaneous, $(1 - N)/K$ would be negative and population size would decline from this point. In this lagged simulation however, the feedback term incorporates the density three time steps previous, at point $t = 7$, giving a value of $(1 - N)/K = (1 - 151/500) = 0.7$, so rapid population growth continues. It is not until point $t = 13$ that the $N(t - 3)$ exceeds K, causing dN/dt to become negative. The lag causes population growth to overshoot the carrying capacity, rising for three time steps to point $t = 13$ before density dependence begins to compensate for the fact that $N > K$, and population size turns downward. When population size falls below K to point $t = 16$, the same thing happens in reverse. N at time $t = 13$ determines the feedback at point $t = 16$, so the population density continues to fall until point $t = 19$, when growth again becomes positive. In Figure 2.7, successive overshoots and undershoots cause an **oscillation**. Because each phase of the oscillation rising above and falling below K takes about τ time steps, its **period** (the time between successive maximum or minimum values) will be roughly 4τ.

Lagged logistic populations do not always oscillate. Recall that populations with a small growth rate take a long time to regain equilibrial density after perturbation. If the growth rate (r) is small and the lag time (τ) is short, the difference between $N(t)$ and $N(t-\tau)$ will be small, and the lag will not have much effect on density-dependent feedback. Under these conditions, a lagged logistic population may simply approach K a little sooner than the instantaneous, continuous logistic of equation (2.2), without overshooting the equilibrial density. The likelihood of oscillation increases with population growth rate, r, and the lag time, τ. For the lagged logistic model given by equation (2.12), specific values of the product $r \bullet \tau$ determine whether and how much oscillation occurs. The e in Table 2.1 is the base of natural logarithms, with a value of approximately 2.72, so $e^{-1} \cong 0.37$, and $\pi/2 \cong 1.57$. A *monotonic approach to equilibrium* rises or falls steadily without oscillation like the sigmoid N vs t trajectory in Figure 2.4a. To run a *Populus* simulation with the lagged logistic model that gives a monotonic approach to equilibrium, set $N(0) = 5$, $K = 500$, $r = 0.1$, and $\tau = 2$. A *stable limit cycle* is an oscillation that persists indefinitely with the same **period** and **amplitude** (the difference between maximum and minimum values). For an example, change the parameter values of the monotonic run to $r = 0.5$, and $\tau = 4$.

Equation (2.12) gives just one of many possible lagged logistic formulations. For another example, feedback might reflect some average of previous population densities, rather than one particular value (May 1976). The exact $r \bullet \tau$ criteria for stable equilibria or cycles would differ for this analog, but its basic dynamics would not change. In particular, the tendency toward oscillation (and the amplitude of those oscillations) would increase with population growth rate and the lag in density-dependent feedback.

Table 2.1

CRITICAL VALUES FOR THE LAGGED LOGISTIC MODEL
THE PRODUCT OF r AND T DETERMINES THE DYNAMICS OF A LAGGED LOGISTIC MODEL.*

$r \bullet \tau$ value	Lagged Logistic Dynamics
$0 < r \bullet \tau < e^{-1}$	Monotonic approach to a stable equilibrium
$e^{-1} < r \bullet \tau < \pi/2$	Oscillations damping to a stable equilibrium (Figure 2.7, where $r \bullet \tau = 1.5$)
$r \bullet \tau > \pi/2$	Stable limit cycles

*For the version given by equation (3.12), (May, 1976).

2.3 DISCRETE LOGISTIC MODELS

A population with discrete generations or cohorts cannot adjust instantaneously to changes in density-dependent feedback because births occur only once in each generation or cohort interval. There is an implicit lag associated with the period of discrete population-growth increments. With the lagged logistic model of the previous section, lag time, τ, could vary in length; but with a discrete logistic model it is constant, fixed by the interval of discrete time steps. As a result, r and K alone determine the dynamics. When r is small, the population may not grow fast enough to overshoot carrying capacity within the lag time of a single cohort interval; but as r increases, sustained oscillations are more likely. *Populus* outputs of discrete logistic dynamics (Figure 2.8) illustrate these general expectations. The input window (Box 2.3) reformulates the list of plot options for use with discrete data.

It is not obvious what difference equation would provide the best discrete analogy for the continuous logistic equation, (2.2), and two methods have been used to establish a correspondence between continuous and discrete models of density-depen-

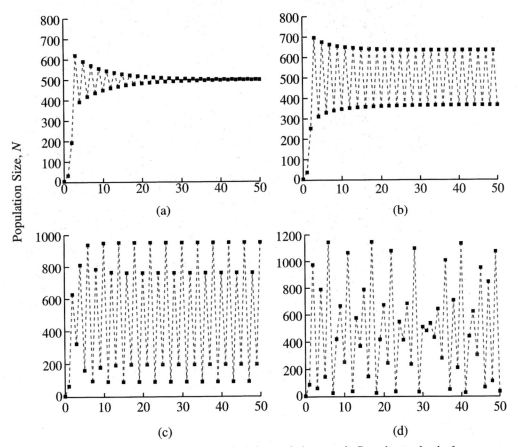

Figure 2.8   *Populus* simulations of discrete logistic population growth. Growth rates for the four cases are (a) $r = 1.9$; (b) $r = 2.05$; (c) $r = 2.6$; and (d) $r = 2.91$. The environmental carrying capacity and initial population sizes were held constant at $K = 500$, and $N_0 = 5$, respectively. (d) shows chaotic cycles, which never settle into a repeating pattern.

BOX 2.3

The *Populus* input window for density-dependent population growth, with parameter values and option settings that produce the simulation of discrete logistic growth in Figure 2.8a.

Choosing the discrete model reconfigures the list of plot types to reflect the finite time steps. Output points are now plotted with dotted connections, rather than the smooth curves indicating a continuous function.

Use the toggle button on the parameter box for the intrinsic rate of increase, *r*, to explore the interaction between growth rate and the complex dynamics of this model.

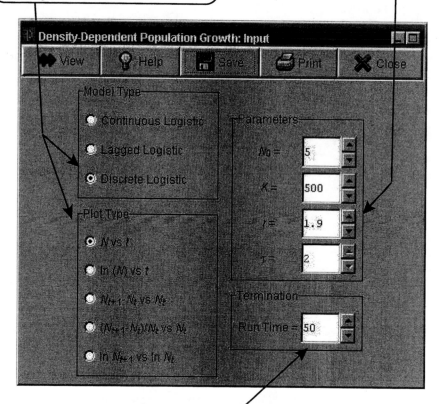

Using *r* values large enough to produce chaotic dynamics simulate some long run times to show that population size never retraces the same cycle.

dent growth (May *et al*. 1974). One method is to choose a difference equation giving a series of finite points along the *N* vs *t* trajectory of continuous logistic population growth (Pielou 1969, Gurney and Nisbet 1998). This approach simply mimics the dynamics of a continuous logistic model; these formulations do not portray the overshoots and oscillations caused by implicit lags in a discrete system. Alternatively, one could choose a difference equation that is equivalent to the continuous logistic at the limit where smaller and smaller time increments become instantaneous, and project this process in discrete time steps. This second approach reveals a rich assortment of dynamic oscillations associated with density-dependent feedback in discrete systems. May and Oster (1976) review several different mathematical formulations of this second type and analyze their properties, which are qualitatively similar. The discrete logistic model implemented for the *Populus* software is a common representative of this second class:

$$N_{t+1} = N_t e^{r\left(1 - \frac{N_t}{K}\right)} \tag{2.13}$$

The model uses an instantaneous growth rate, *r*, rather than a finite growth rate, λ, because of its derivation as an instantaneous equivalent of the continuous logistic model.

DYNAMICS AND EQUILIBRIA

It is easy to see that equation (2.13) causes population-growth increments, $N_{t+1} - N_t$, to fall as *N* approaches *K* and the feedback term approaches 0, because $e^0 = 1$. With a small population-growth rate, *r*, this discrete model gives a sigmoid approach to equilibrium, just like continuous and lagged logistic models. With increasing *r* values, discrete logistic dynamics show damped oscillation (Figure 2.8a); then 2-point cycles of constant period and amplitude (Figure 2.8b); and then cycles that include 4 points (Figure 2.8c), 8 points, 16 points, etc., before repeating. Finally, very large *r* values cause population size to fluctuate in a way that is extremely sensitive to initial conditions, and never settles into a precisely repeating cycle (Figure 2.8d), a regime that mathematicians term chaotic. For the discrete logistic defined by equation (2.13), growth rates that bound these different dynamic behaviors are given in Table 2.2 (May 1976).

Table 2.2

CRITICAL VALUES FOR THE DISCRETE LOGISTIC MODEL
GROWTH RATE DETERMINES WHETHER A POPULATION WITH DISCRETE LOGISTIC GROWTH WILL OVER-SHOOT AND OSCILLATE WITHIN THE FIXED LAG TIME SET BY GENERATION OR COHORT INTERVALS.

Growth Rate, *r*	Discrete Logistic Dynamics
$0 < r < 2.0$	Stable equilibrium point
$2.0 < r < 2.526$	2-point cycle
$2.526 < r < 2.656$	4-point cycle
$2.656 < r < 2.685$	8-point cycle
$2.685 < r < 2.692$	16-, 32-, 64-point cycles, etc.
$r > 2.692$	Chaos

Redrawn after May, 1976.

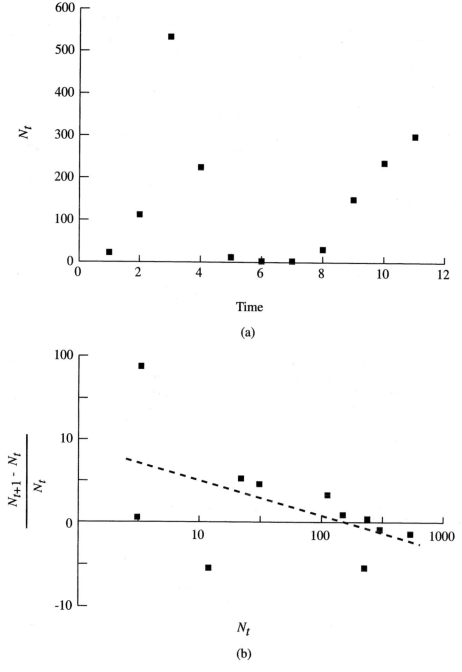

Figure 2.9 Empirical data on black-headed budworm abundance from New Brunswick (Morris 1959), illustrating a time series of forest-insect densities over eleven successive years. (a) shows a time trajectory, N vs t, of population densities, with no apparent pattern. In (b), *per capita* changes are plotted against population density on log–log scales. The dotted line gives a least-squares regression through these data; it suggests a negative trend in *per capita* population growth as density increases but is not statistically significant. (Figure redrawn after Morris 1959)

2.4 DETECTING DENSITY DEPENDENCE

That a relatively simple equation like (2.13) can give the complex fluctuations of Figure 2.8 raises an interesting question. Populations that grow exponentially, without any density-dependent regulation, might fluctuate around some long-term average density in a random succession of good- and bad-weather years. If lagged and discrete logistic populations also fluctuate, how can we determine whether density-dependent regulation is occurring? The answer is that while logistic growth may be chaotic, it is not **random**. In a regulated population, we expect *per capita* growth from N_t to N_{t+1} to vary inversely with N_t. Thus a graph of $(N_{t+1} - N_t)/N_t$ vs Nt for a population with density-dependent regulation should show the declining form of Figure 2.2. Figure 2.9 gives a historic example from the work of Canadian entomologist R. F. Morris (1959), who produced a **time series** of density estimates over 12 black-headed budworm generations from 1946 to 1958 in New Brunswick forests. While the N vs t plot (Figure 2.9a) shows no clear pattern, regression through values of $(N_{t+1} - N_t)/N_t$ plotted against N_t gives a negative trend, suggesting density-dependent regulation. Morris proposed a second graphical analysis of these time-series data, illustrated in Figure 2.10. Here consecutive population sizes are plotted as N_{t+1} vs N_t on log–log axes. The regression through these points has a slope less than 1.0, implying that the growth increment following high densities is smaller than the growth increment following low densities.

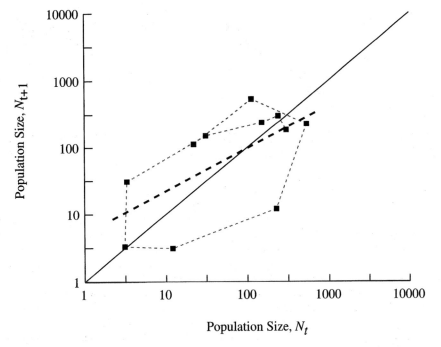

Figure 2.10   Log–log plot of successive budworm densities. These are the same Morris (1959) data from Figure 2.9. Each point on this graph represents a consecutive pair of population sizes, with log N_t on the *x*-axis, and log N_{t+1} on the *y*-axis. The dotted line gives a least-squares regression through these data; the regression slope less than 1.0 suggests a negative trend in *per capita* population growth as density increases but is not statistically significant.

It is telling that neither of the regressions in Figures 2.9 and 2.10 quite reaches statistical significance. In fact, it is difficult to detect density-dependent regulation in empirical data, particularly with the Morris plot technique of Figure 2.10. Figure 2.11 plots four different *Populus* simulations to illustrate the problem. Figure 2.11a shows the density-independent null expectation, simulating discrete, geometric growth with $\lambda = 3$. Figure 2.11b illustrates continuous logistic growth with a small growth rate, $r = 0.3$. The simula-

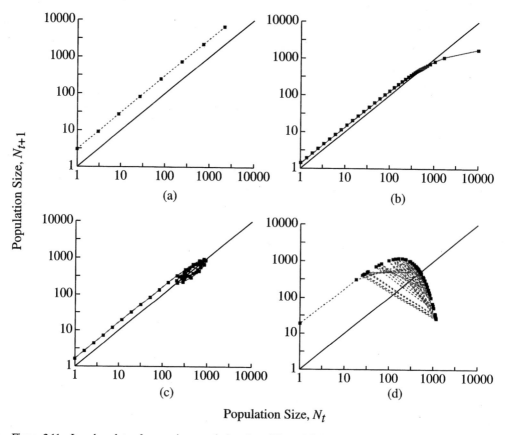

Figure 2.11   Log–log plots of successive population sizes (N_{t+1} vs N_t) from time series produced by four *Populus* simulations of density-independent and density-dependent population growth. Each of the four graphs also has a 45° reference line, $x = y$. Where simulation plots cross this line, $N_t = N_{t+1}$, and $dN/dt = 0$. (a) plots discrete geometric (density-independent) growth with $\lambda = 3$, and $N_0 = 1$; note that it runs above the 45° reference line because each successive population size is λ times larger than its predecessor. It is parallel to the reference line because the *per capita* growth increment does not change with population size in this density-independent simulation. (b) combines two logistic simulations with smaller growth rates ($r = 0.3$). One begins at $N_0 = 1$ and progresses to the right, growing to equilibrium at $N = K = 500$. The other begins at $N_0 = 10,000$ and falls to the same equilibrium, running from right to left on the graph. (c) gives the log–log plot of N_{t+1} vs N_t for the lagged logistic example in Figure 2.7. The growth rate for this simulation was $r = 0.5$, with a time lag of $\tau = 3$ time steps. (d) presents the time series resulting from a discrete logistic simulation with $r = 2.91$. Successive population sizes in this run result from an oscillation like the one shown in Figure 2.8d but with a starting value of $N_0 = 1$.

tion trajectory lies close to the 45° reference, and the slope declines only slightly to equilibrium at $N = K$. This decline in slope is so subtle that with noisy empirical data it would be difficult or impossible to detect density-dependent regulation using this technique. Because each datum is plotted on both the x and y axes, measurement errors also reduce the N_{t+1}, N_t correlation, lowering the regression slope as a statistical artifact (Case 2000). The same problem appears in lagged simulations with modest growth rates (Figure 2.11c). It is only with high growth rates (Figure 2.11d) that the Morris plot yields a dramatic result, and even this can be difficult to distinguish from stochastic noise (Dennis and Taper 1994, Ives 1995).

Density dependence is conceptually important because it regulates population size. It is ironic that the dynamics of many natural populations must be classified as density vague (Strong 1986) because we can not certify regulation or its absence convincingly. The existence of density-dependent effects does not preclude the concurrent action of density-independent and stochastic phenomena, and the effort to disentangle these processes continues. Students who would like to pursue these issues in depth can enter the recent literature through Dennis and Taper (1994) and Ives (1995).

REFERENCES

Case, T. J., *An Illustrated Guide to Theoretical Ecology*. New York: Oxford University Press, 2000, pp. 103–155.

Davidson, J. and H. G. Andrewartha, "Annual Trends in a Natural Population of *Thrips imaginis* (Thysanoptera)," *J. Anim. Ecol.*, vol. 17, pp. 193–199.

Davidson, J. and H. G. Andrewartha, "The Influence of Rainfall, Evaporation and Atmospheric Temperature on Fluctuations in the Size of a Natural Population of *Thrips imaginis* (Thysanoptera)," *J. Anim. Ecol.*, vol. 17, pp. 200–222.

Dennis, B. and M. L. Taper, "Density Dependence in Time Series Observations of Natural Populations: Estimation and Testing," *Ecol. Monogr.*, vol. 64, pp. 205–224.

Gurney, W. S. C. and R. M. Nisbet, *Ecological Dynamics*, New York: Oxford University Press, 1998, pp. 61–64.

Hutchinson, G. E., *An Introduction to Population Biology*, Yale University Press, 1978, pp. 1–40.

Ives, A. R., "Measuring Resilience in Stochastic Systems," *Ecol. Monogr.*, vol. 65, pp. 217–233.

May, R. M., "Biological Populations with Non-overlapping Generations: Stable Points, Stable Cycles, and Chaos," *Science*, vol. 156, pp. 645–647.

May, R. M., "Models for Single Populations." In: *Theoretical Ecology: Principles and Applications*, (R. M. May, ed.), Sunderland, MA, Sinauer Associates, 1976.

May, R. M., G. R. Conway, M. P. Hassell and T. R. E. Southwood, "Time Delays, Density Dependence and Single-species Oscillations," *J. Anim. Ecol.*, vol. 43, pp. 747–770.

May, R. M. and G. F. Oster, "Bifurcations and Dynamic Complexity in Simple Ecological Models," *Am. Nat.* vol. 110, pp. 573–599.

Morris, R. F., "Single Factor Analysis in Population Dynamics," *Ecology*, vol. 40, pp. 580–588.

Neuhauser, C., *Calculus for Biology and Medicine*. Upper Saddle River, NJ: Prentice Hall, 2000, pp. 355–366.

Nixon, C. M., "White-tailed Deer Growth and Productivity in Eastern Ohio," *Game Res. Ohio*, vol. 3, pp. 123–136.

Pielou, E. C., *An Introduction to Mathematical Ecology*, New York: John Wiley & Sons, 1969, pp. 19–32.

Pimm, S. L., *The Balance of Nature? Ecological Issues in the Conservation of Species and Communities*, Chicago, University of Chicago Press, 1991, pp. 18–34.

Ricklefs, R. E. and G. L. Miller, *Ecology* (4th ed.), New York: W. H. Freeman and Co., 1999, pp. 314–328, 346–359.

Roughgarden, J., *Theory of Population Genetics and Evolutionary Ecology: An Introduction*, New York: MacMillan-Collier, 1979, pp. 303–310.

Smith, F. E. "Density Dependence in the Australian Thrips," *Ecology*, vol. 42, pp. 403–407.

Strong, D. R. "Density Vagueness: Abiding the Variance in the Demography of Real Populations." In: *Community Ecology*, (J. Diamond and T. J. Case, eds.), Harper & Row, 1986.

Tanner, J. T., "Effects of Population Density on Growth Rates of Animal Populations." *Ecology*, vol. 47, pp. 733–45.

PROBLEMS AND EXERCISES

CONTINUOUS LOGISTIC POPULATION GROWTH

1.   Investigate the effect of initial population size on continuous logistic population growth. Use *Populus* to simulate population growth when $N(0) = K$, $N(0) < K$, $N(0) > K$. Describe the curves that result from these simulations for each of the continuous logistic plot types listed in Box 2.1.

2.   Intuitively, how should increasing r affect logistic population growth? Test the prediction you made with *Populus* simulations, running continuous logistic cases with $r = 0, 0.5$, and 5, holding $N(0)$ and K constant.

3.   Run a *Populus* simulation of continuous, logistic population growth using the default parameter values which are $N(0) = 5, K = 500$, and $r = 0.2$. Using the plotting option that grids output graphs, count the number of time steps required for the population to grow from $N = 5$ to $N > 475$. Call this interval the population's *return time* to $N > 475$ after a perturbation to $N = 5$. Using the same $N(0)$ and K values, how would you expect the return time to change if r is doubled to $r = 0.4$? What if it is halved to $r = 0.1$? Check your expectation by running the appropriate simulations. Investigate similar cases beginning with $N(0) > K$, so that the return toward equilibrium involves a decline in population size. Does the return from $N(0) = 995$ to $N < 525$ happen faster, slower, or at the same speed, relative to the return from $N(0) = 5$ to $N > 475$ with the same r value? Why?

4.   How quickly will a logistic population double from the default conditions, $N(0) = 5, K = 500$, and $r = 0.2$? Can you derive an expression for logistic doubling time without looking back at equations (2.8) through (2.11)?

LAGGED LOGISTIC GROWTH

5.   Use *Populus* simulations to compare the continuous logistic model with a lagged logistic model; set $\tau = 0$, but otherwise use the default values, $N(0) = 5$, $r = 0.2, K = 500$. Do you expect the two plots to differ? Check your prediction. Then set $\tau = 1$; how and why does this plot differ from the plot with $\tau = 0$? Try values of $\tau = 3$, and $\tau = 5$. What is the effect of increasing the lag time? Why?

6.   Reset to the default values of $\tau = 2$ and $r = 0.2$. Run this simulation, and then gradually increase r using the toggle button, checking values up to $r = 1.5$. After the population begins to cycle, what happens to the period of the cycles as r increases? What happens to the amplitude? What parameters affect the period and amplitude?

DISCRETE LOGISTIC GROWTH

7.   How do variations in r affect the dynamics of the discrete logistic model? Use default values of $N(0)$ and K, and vary r using the following values: $r = 0.5, 1.75$, 2.0, 2.3, 2.6, 2.65, 2.7, 2.75, 3, and 5. Can you find an r value that divides repeating cycles from chaos? Why does changing the intrinsic growth rate have these effects? Can you suggest examples of real populations where these effects might be important?

PROBLEMS WITH EMPIRICAL DATA

8.   The final problem in Chapter 1 on density-independent population growth presented a table of worldwide human population estimates for the period from 1950 to 1998. If you have not worked through that problem, go back and do so now. You will find that through much of the latter twentieth century, human population growth has been faster than exponential. Predict *a priori* and explain the expected shape of a plot showing log N_{t+1} vs log N_t for those data. What will the slope of the plot be during the early and late portions of this growth interval? Does it, or will it ever, cross the $y = x$ line?

9.   Ecology students at the University of Minnesota often run a lab experiment on the growth of bacterial cultures. They inoculate sterile growth medium in a side-arm culture flask with *Escherischia coli*, and collect absorbency data with a spectrophotometer as increasing bacterial densities make the incubated culture more turbid. Data in the following table give results from two such experiments, one with Low and one with High initial nutrient concentrations in the culture medium. Do these data suggest the action of density-dependent processes?

	Absorbency	
Minutes	Low	High
0	0.1	0.08
15	0.14	0.14
30	0.16	0.21
45	0.2	0.31
60	0.29	0.45
75	0.33	0.55
90	0.42	0.7
105	0.62	1.1
120	0.69	1.3
135	0.75	1.6
150	0.8	1.8
165	0.85	1.8
180	0.88	1.9

Try the following analysis:

(a) Make a spreadsheet for these data using a program like Excel or Lotus 123. This will facilitate numerical and graphical manipulations.

(b) Absorbency is only an analog of the bacterial population size, so the actual numbers are not important. Transform the absorbency data, multiplying the numbers in each column by 100 to produce a series of positive whole numbers (integers) that will be easier to think of as population sizes and more convenient. They will range from 10 to 88, and 8 to 199 for the low- and high-nutrient cultures, respectively.

(c) For each data set, make graphs of the following: N vs t, log N vs t, $(N_{t+1} - N_t)$ vs N_t, $(N_{t+1} - N_t)/N_t$ vs N_t, and log N_{t+1} vs log N_t. For each graph, explain (*i*) how a density-independent population is expected to look, (*ii*) how a density-dependent population is expected to look, and (*iii*) whether the empirical data from bacterial cultures more closely resemble the density-dependent or density-independent expectation.

(d) Next, use your graphs to make estimates of r and K. Which graphs (or portions of graphs) give the best evidence of these important logistic parameter values? Why? Be sure to specify the units in which your r value is measured.

(e) Use your empirically derived r and K values to run a *Populus* simulation of continuous logistic population growth. Print a copy of the N vs t and $(N_{t+1} - N_t)/N_t$ vs N_t output graphs, and plot the lab data on bacterial growth directly on those graphs. How well do the empirical data conform to the theoretical expectation?

(f) What aspects of life in a culture flask might slow bacterial growth after three hours of incubation? Again, discuss this question.

10.   Davidson and Andrewartha (1948a, Table 3) presented a time series of density estimates for thrips (very small insects) infesting Australian roses for 81 months between 1932 and 1938. Obtain a copy of their paper from your

library and apply the analyses of problem 9 to those data to determine whether they show evidence of density-dependent effects. After making this analysis, read the Davidson and Andrewartha paper and the following paper in which they give interpretations (1948b). After deciding whether you agree with Davidson and Andrewartha, read the reanalysis of their data by Smith (1961). Finally, decide for yourself; do these data suggest the action of density-dependent processes?

FOR ADVANCED STUDENTS

11.   Set the run-time for *Populus* simulations of discrete logistic population growth to 24, select the log (N_{t+1}) vs log (N_t) output, and run several simulations, beginning first with $r = 0.05$, and subsequently using larger values up to $r = 3$. Compare the plot to a reference line $x = y$, like those plotted in Figure 2.11. Which r values cause the plot and reference line to differ most strongly? Does the starting density, N_0, have any effect on your ability to distinguish the simulation plot and reference line? Discuss these results with your class. Is the log–log plot of subsequent population densities a valid technique for detecting density dependence?

12.   At the beginning of this chapter, we defined **density dependence** in a way that implied population growth rate to be a **function** of population size, N.

$$\frac{dN}{dt} = f(N)$$
(2.14)

If you recall Taylor's Theorem from your math training, you will remember that this function can be expanded as a **power series**,

$$\frac{dN}{dt} = a + bN + cN^2 + dN^3 + eN^4 + \ldots$$
(2.15)

where a, b, c, etc. are constants that specify the form of our function. The first term in this series (a which equals aN^0 and is called the zero-order term because of this 0 exponent) is not useful in population models. A population does not exist if $N = 0$, and it contains no members, so $a = 0$. If we take the next simplest case,

$$\frac{dN}{dt} = bN$$
(2.16)

where $a, c, d, e, \ldots = 0$, we have a first-order equation (because $N = N^1$) that you should recognize as the density-independent exponential growth model (b is the constant intrinsic growth rate, r). In a second-order power series, b and c have non-zero values, giving

$$\frac{dN}{dt} = bN + cN^2$$

(2.17)

a.   Can you show that the continuous logistic model is a second-order equation, algebraically equivalent to equation (2.17)? If you need a hint, set $c = -r/K$.

b.   Now return to examine Figure 2.2. The linear feedback function (Fig. 2.2a) comes directly from the logistic model, as described in the text. How many power series terms would be necessary to specify the feedback functions given by curves Figures 2.2b and 2.2c? Can you find values for the power series constants that will produce these curves? Experiment a little with this problem, using a spreadsheet like Excel or Lotus 123. Set a column of values ranging from 0 to 100 in increments of 1, and call these your N values. Now you can specify a formula to give a column of dN/Ndt as a function of the adjacent N's. By trying different values of the power series constants, can you find values giving curvilinear functions of dN/Ndt on N that range from 0.2 at $N = 0$ to 0 at $N = 100$? With some experimentation you may be able to reproduce the curves presented in Figure 2.2b & c exactly .

CHAPTER 3

Demography and Age-Structured Population Growth

The simplicity of population-growth models in Chapters 1 and 2 rests partly on their assumption that all individuals have identical probabilities of dying or producing offspring. This assumption is often incorrect. Among humans, for example, fertility peaks early in the third decade of life, while mortality declines from juvenile age to reproductive maturity and then accelerates in post-reproductive years. This implies that population-growth rates may depend on **age structure**, the population's composition with respect to age. In this chapter, we will examine some basic issues in **demography**, the study of population age structure, and explore their effects on population dynamics.

Time and age are continuous variables, but we often think of age in discrete units. If someone asks how old you are, you probably answer with the number of birthdays you've celebrated. Demographers also divide the life span into a series of discrete intervals, each representing a **cohort** of individuals who are approximately the same age, and tabulate the average probabilities of reproduction and survival for each interval. Then after counting the proportion of the population in each cohort, they can project population growth with weighted averages of the fertility and survival rates over all age intervals.

If births and deaths occur continuously, we can divide the life span of organisms into any arbitrary number of discrete age classes to make a simple or a detailed demographic analysis. Often, however, births and deaths are seasonal, concentrated in discrete pulses. Consider the North American white-tailed deer, *Odocoileus virginianus*. A cohort of new fawns is born each spring, and hunters harvest about one third of the population each fall. It makes sense to let this annual cycle of births and deaths set the interval of periodic analyses, and we could keep track of population size with a census taken immediately after each new cohort appears. Figure 3.1 illustrates the sequence of events over three successive years. Newborns that appear in the first census are only a few days or weeks old, and we call this cohort the "zero" age class. At the second census, members of this same cohort have survived to the "yearling" or one-year-old age class. Female deer may produce their first offspring as yearlings. These fawns appear as newborns with their one-year-old mothers in the second census. Each subsequent census will include a cohort of newborns, yearlings, two-year-olds, three-year-olds, etc. that are surviving members of the cohorts born one, two, and three years earlier. All cohorts except the newborns may contribute to the newborn fawn crop that appears for the first time in that same census.

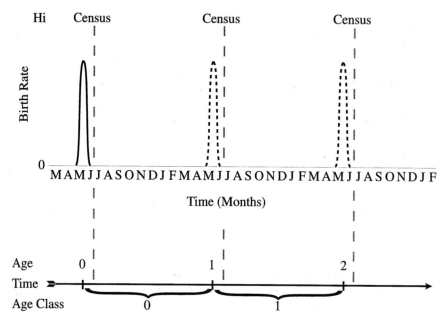

Figure 3.1 Schematic life history for white-tailed deer, *Odocoileus virginianus*. In Minnesota the annual pulse of newborn fawns appears in May. Letters on the *x*-axis represent months of the year. If we follow the dynamics of deer with an annual post-reproductive census, each count will include a series of discrete cohorts whose members are all about the same age. The time line in the bottom half of the figure tracks the age of the first cohort to appear. These deer are tabulated in their first census as newborn fawns. They must survive until the following May (with probability l_1) to produce offspring of their own (whose expected number is m_1). Subsequently these deer appear in their second census as yearlings.

3.1 ESTIMATING THE SURVIVAL AND FERTILITY PARAMETERS S_x, l_x, m_x AND F_x

McCullough (1979) studied white-tailed deer at the George Reserve in southeastern Michigan from 1962 until 1974. Almost all mortality in this population resulted from autumn hunting that held densities to 47–73 females (does) and 30–54 males (bucks) in the 2-square-mile fenced preserve. Data on the sex and age of each harvested deer allowed McCullough to reconstruct the age-composition of this population. Table 3.1 illustrates his reconstruction for females, giving S_x, the number of survivors annually for each age, x. Demographers studying polygynous species (where few males inseminate many females) often focus exclusively on females because female survival and fertility govern the population dynamics. The age-specific survivorship, l_x, is an individual's probability of surviving from birth to age x. From the S_x values of Table 3.1 we can calculate l_x as S_x divided by the S_0 value (i.e., the number of newborns) observed x years earlier. By definition, l_0, the probability of surviving from birth to birth, is 1. From Table 3.1, the l_1 estimates (probabilities of surviving from birth to age 1) for doe fawns born in 1966–1969 are 20/23, 10/18, 13/25, and 11/17, respectively. For all twelve years of the study combined, the average probability that a new fawn would be alive one year later was ≈ 0.713 (the symbol ≈ means "approximately equal to"). We list the l_x esti-

Table 3.1

A SAMPLE OF MCCULLOUGH'S DATA ON THE AGE STRUCTURE OF GEORGE RESERVE DEER.

x	1966	1967	1968	1969	1970
0	23	18	25	17	23
1	14	20	10	13	11
2	15	9	16	8	7
3	9	10	7	5	4
4	4	6	4	2	3
5	5	3	2	3	
6		5	3	2	1
7	1		4	3	2
8	1			2	2
9	1	1			1
10					
11	1				

These data are reproduced by permission of Dr. Dale R. McCullough and the University of Michigan Press. Values in the table give S_x, the number of surviving females in each age class over a five-year portion of the study. It was possible to estimate S_x retrospectively because the age of each individual was determined by counting the annuli of cementum in the roots of its teeth when it died.

mates for all age classes in a **life table**, or age-specific schedule of survival and fertility (Table 3.2). In an analogous set of l_x estimates for male deer, we would find age-specific survivorship somewhat lower. Both the mate-seeking behavior of bucks and the preference of hunters for deer with antlers make the bucks more vulnerable than does.

Table 3.2

LIFE TABLE FOR THE WHITE-TAILED DEER AT MICHIGAN'S GEORGE RESERVE FROM 1964–1972.

x	l_x	m_x
0	1.000	0.000
1	0.713	0.216
2	0.460	0.537
3	0.298	0.874
4	0.167	0.795
5	0.094	0.843
6	0.073	1.150
7	0.055	0.460
8	0.024	0.805
9	0.010	0.920
10	0.005	0.000

The l_x estimates are those of Mccullough (1979); m_x estimates combine fertilities from two periods with different densities, yielding an analysis that is slightly simplified (for teaching purposes) from McCullough's original version. These data are reproduced by permission of Dr. Dale R. McCullough and the University of Michigan Press.

Age-specific fertility, m_x, is the expected number of female births per female during the interval from age $x - 1$ to age x. Newborns do not produce progeny and m_0 is undefined, so m_1, the *per capita* expectation of female progeny during the interval from age 0 to age 1, is the first non-zero fertility in a life table. McCullough was able to estimate the proportion of George Reserve does in each age class carrying one, two, or three embryos by dissecting the reproductive tract of each harvested deer. Almost half (47%) of the yearlings bred, producing a single fawn. The sex ratio at birth was about 46% female, so m_1, the average number of doe fawns born *per capita* to yearling mothers is the product of the fraction of does breeding (0.47), the average number of embryos per doe (in yearlings, 1.0), and the sex ratio, or proportion of fawns that are female (0.46):

$$m_1 = (0.47)(1)(0.46) = 0.216 \tag{3.1}$$

All of the 2-year-old does bred, producing an average of 1.17 fawns *per capita*, so

$$m_2 = (1)(1.17)(0.46) = 0.537 \tag{3.2}$$

Fertility continued to increase with the age and social rank of does, reaching a peak at $m_6 = 1.15$. Again, because of the male-biased birth sex ratio, this means that 6-year-old does produced an average of 2.5 fawns. Figure 3.2 shows *Populus* plots of the l_x and m_x values from Table 3.2. The m_x estimates for ages 5–9 are variable, influenced more strongly that those for ages 1–4 by sampling error, due to the limited number of older deer harvested from the George Reserve.

We defined two measures of survival, S_x the number of survivors in a cohort of age x, and l_x, an individual's probability of surviving from birth to age x. The age-specific fertility, m_x is a *per capita* expectation, analogous to the individual survival parameter l_x. We can also define a fertility parameter that characterizes the performance of a full cohort, analogous to the total number of cohort survivors, S_x. F_x is the number of female offspring produced by the entire female cohort of age x. F_x is directly observable in a population where we know the age of every individual and can associate offspring with their mothers. From McCullough's data, we can only estimate F_x as the product of S_x and m_x, rounded to integer values in Table 3.3. The accuracy of these estimates, shown by comparing the sum of F_x values at the bottom of each column with S_0 (Table 3.1), varies from year to year. Factors contributing to this variation include year-to-year differences in the number of older individuals in the population, the age-specific accuracy of m_x estimates, and conditions of population density, climate, and resource quality that affect embryo survival.

There is some duplication in defining two measures of both survival (S_x and l_x) and fertility (m_x and F_x). The benefit is that some demographic principals are more simply stated using cohort measures, and some are easier with individual measures. It may help with the cost of remembering four definitions to note that S_x and F_x, with upper-case labels, refer to the cohort parameters, while l_x and m_x, with lower-case labels, measure the survival and fertility of individuals.

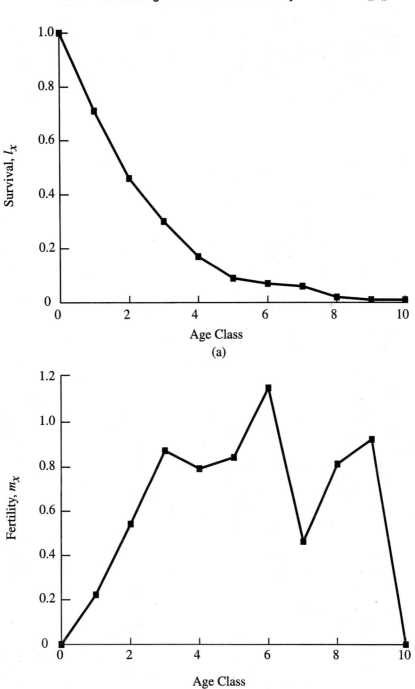

Figure 3.2   *Populus* plots of age-specific survival (a) and fertility (b) for white-tailed deer, *Odocoileus virginianus*, at Michigan's George Reserve, 1962–1974 (McCullough 1979). These graphs illustrate the $l_x m_x$ schedule of Table 3.2. Fertilities were estimated by dissecting embryos from the reproductive tracts of harvested deer. Because relatively few individuals reached advanced age, there is more measurement error associated with the m_x estimates for older cohorts.

Table 3.3

F_x ESTIMATES FOR THE GEORGE RESERVE DEER

x	1966	1967	1968	1969	1970
0	0	0	0	0	0
1	3	4	2	3	2
2	8	5	9	4	4
3	8	9	6	4	3
4	3	5	3	2	2
5	4	3	2	3	0
6	0	6	3	2	1
7	0	0	2	1	1
8	1	0	0	2	2
9	1	0	0	1	
Σ	28	32	27	21	17

Made by multiplying S_x (Table 3.1) and m_x values (from Table 3.2).

3.2 POPULATION PROJECTION FROM THE LIFE TABLE

McCullough was able to infer the age structure of the George Reserve deer population retrospectively by knowing their ages at death. In a similar way, we can project population size and age structure into the future, using the age-specific schedule of survival and reproduction (l_x and m_x) given in the life table. Table 3.4 gives an example of this procedure. The first three columns reproduce the $l_x m_x$ schedule of Table 3.2. The fourth column, labeled $S_x(0)$, gives an initial population composition for this hypothetical example. In this case, we begin with a population of 24 newborn fawns and assume that each annual population census occurs immediately after reproduction, as illustrated in Figure 3.1. This means that the fawns born just prior to the census at time $t(0)$ must survive a full year before they reproduce. Note that for purposes of illustration, we will assume that individuals survive independently. In reality, a fawn could probably not survive without its mother before reaching 4 or 5 months' age.

To project the population composition expected in a census at time $t(1)$, we first determine how many of the $t(0)$ fawns will survive to $t(1)$, multiplying 24 by l_1, or $24 \times 0.713 = 17.1$. This is the expectation for $S_1(1)$. While deer only come in whole integer numbers, the expected number can be between 17 and 18. These S_1 expected survivors will produce $S_1 \times m_1$ progeny that will appear with them in the census at $t(1)$, so the expectation of newborn fawns at time $t(1)$, $S_0(1)$ is $17.1 \times 0.216 = 3.7$. Now projecting forward to time $t(2)$, the expected number of surviving two-year-old individuals is the number of newborns that appeared two years earlier at $t(0)$, multiplied by l_2 ($24 \times 0.46 = 11$). The expected number of surviving one-year-old individuals is the number of newborns that appeared one year earlier at $t(1)$, multiplied by l_1 ($3.7 \times 0.713 = 2.6$). These products give $S_2(2)$ and $S_1(2)$. Multiplying them by m_2 and m_1, respectively, gives the F_2 and F_1 expectations at time $t(2)$, and the sum of these F_1 and F_2 values then gives the expected $S_0(2)$.

Using the technique of Table 3.4, we can project population composition forward indefinitely. The "tabular projection" output option in the *Populus* Age-Structured

Table 3.4

PROJECTIONS OF $S_x(T)$, THE NUMBER OF INDIVIDUALS EXPECTED TO SURVIVE, BY AGE, FROM THE DEER LIFE TABLE.

x	m_x	l_x	$S_x(0)$	$S_x(1)$	$S_x(2)$	$S_x(3)$	$S_x(4)$	$S_x(5)$	$S_x(6)$
0	0.000	1.000	24.000	3.700	6.495	8.165	7.010	7.182	8.134
1	0.216	0.713	0	17.112	2.638	4.631	5.822	4.998	5.120
2	0.537	0.460	0	0	11.040	1.702	2.988	3.756	3.225
3	0.874	0.298	0	0	0	7.152	1.102	1.936	2.433
4	0.795	0.167	0	0	0	0	4.008	0.618	1.085
5	0.843	0.094	0	0	0	0	0	2.256	0.348
6	1.150	0.073	0	0	0	0	0	0	1.752
7	0.460	0.055	0	0	0	0	0	0	0
8	0.805	0.024	0	0	0	0	0	0	0
9	0.920	0.010	0	0	0	0	0	0	0
10	0	0.005	0	0	0	0	0	0	0
		ΣS_x	24.000	20.812	20.173	21.650	20.930	20.745	22.097
		λ		0.867	0.969	1.073	0.967	0.991	1.065
		$S_0/\Sigma S_x$	1.000	0.178	0.322 0	377	0.335	0.346	0.368
		$S_1/\Sigma S_x$	0.000	0.822	0.131	0.214	0.278	0.241	0.232
		$S_2/\Sigma S_x$	0.000	0.000	0.547	0.079	0.143	0.181	0.146

This example begins arbitrarily with 24 newborn females at time zero, and projects the size and composition of the population forward in time for six years based on the life table's schedule of age-specific survival and reproduction. For each year step, we first calculate S_1–S_9 as $S_0(t-x)$ multiplied by l_x. Then, the new S_1–S_9 expectations are each multiplied by the appropriate m_x, and the sum of these products gives S_0. Rows at the bottom of each annual column list ΣS_x, the sum of survivors in all age classes; λ, the discrete growth rate from t-1 to t; and the fraction of survivors in the newborn, yearling, and two-year-old age classes, calculated to show convergence to a stable age distribution.

Population Growth module automates this process for any run time you choose (see the input window illustrated in Box 3.1). Each time step in the projection requires the same pattern of calculations. First, we determine the numbers of survivors in age classes S_1 and older, and then we determine the number of new progeny that those survivors are expected to produce. It is important to realize that this order of calculations (survival first, then reproduction) is dictated by the post-reproductive timing of our censuses, illustrated in Figure 3.1.

It would be equally possible to follow populations with a pre-reproductive census, and in practice the choice is one of convenience; we count the point in a life cycle that is easiest to enumerate. With the pre-reproductive censuses illustrated in Figure 3.3, progeny join the population immediately after one census, must survive a cohort interval before appearing in the next census, and then reproduce. With this alternative approach, the order of computations used to project population age structure would reverse. First we would estimate the reproductive contribution of individuals present at the preceding census, and then we would determine the expected numbers of survivors (in all age classes) at the next census. This difference in the sequence of calculations causes projections based on pre- and post-reproductive sampling to differ, so it is necessary to understand the assumed basis of one's procedure. A life table based on post-reproductive sampling is recognizable because it has an initial l_x of $l_0 = 1$, meaning

BOX 3.1

Populus input window for Age-Structured Population Growth. The data required to run an age-structured simulation are a schedule of l_x and m_x values and a set of initial cohort sizes, $S_x(0)$, which are entered in the scrolling box at the bottom of the window.

The Age-Structured Population module offers 11 different views of demographic parameter changes with both time and age. It is instructive to work through each graph in turn for a single schedule of survival and fertility.

Setting the number of age classes adjusts the size of the scrolling parameter input box at the bottom of the window.

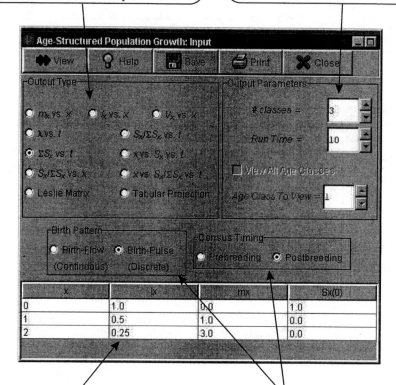

To run an age-structured simulation, you enter a schedule of l_x and m_x values, and an initial population of S_x values. The default values are those used in problem 1 at the end of the chapter.

Note the options for continuous vs discrete reproduction, and pre-breeding vs post-breeding census timing. These choices affect the demographic computations carried out by the program, as described in the text.

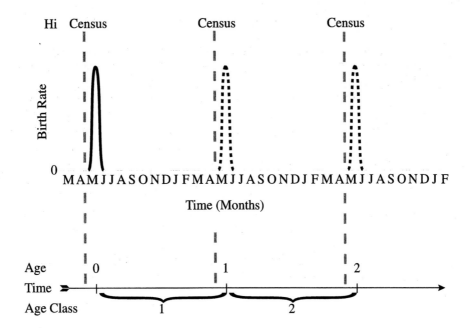

Figure 3.3   A second schematic of the life history for white-tailed deer, *Odocoileus virginianus*. In this case, the timing of census counts is shifted to precede reproduction. The first-born cohort in this figure does not appear in a census until it is nearly a full year old, by which time a fraction of the original cohort members will have perished. The timing of events after the census is reversed relative to the example in Figure 3.1. First, newborns appear, then the population must survive a long interval to reach the next census. The order of calculations required to project the population dynamics is similarly reversed relative to the example in Figure 3.1.

that survival from birth to birth is perfect. With pre-reproductive sampling, there is a period of time and mortality between birth and a cohort's first census appearance. In this case, the initial l_x will be $l_1 < 1$, and the youngest projected age class is S_1. The presence of a newborn S_0 cohort in each post-reproductive projection makes it the more intuitive and commonly used technique. We will return to post-reproductive sampling for population-growth estimates in the next section, leaving further comparison of the two approaches as an exercise at the end of the chapter. The *Populus* input window for Age-Structured Population Growth (Box 3.1) offers both pre- and post-reproductive census options.

3.3   CONSTANT $l_x\, m_x$ SCHEDULES

The projections of Table 3.4 assume that age-specific survival and fertility (l_x and m_x) remain constant. This is not necessarily so. A look at the gravestone dates in any cemetery will demonstrate that our 18th and 19th century ancestors suffered much higher infant mortality than we do now. Thus a life table illustrates population performance at a particular place and time and may not apply generally. Nevertheless, when survival and fertility rates do remain constant for any extended period, there are interesting consequences.

Recall from Chapter 1 that λ is a growth parameter quantifying discrete changes in population size defined as $\lambda = (N_t/N_{t-1})$. When $\lambda > 1$, population size is increasing; when $\lambda < 1$, population size is decreasing; and when $\lambda = 1$, population size is constant. We can calculate λ for each of the projection steps in Table 3.4 as $\Sigma S_x(t)/\Sigma S_x(t-1)$, where the upper-case sigma represents a sum over all age classes, x; thus, $\Sigma S_x(t)$ is the sum of all survivors, or total population size, at time t. Figure 3.4 gives a *Populus* plot of λ values for 15 annual iterations of the Table 3.4 projection. The initial population for this hypothetical example included only newborn fawns, which contribute relatively little reproduction to the next census as yearlings. The population shrinks initially, then grows as the survivors reach their mature fertility at age 3. Ultimately, this varying growth rate converges to a constant value of $\lambda \approx 1$. Thus shooting at the George Reserve maintained relatively constant deer densities. With less shooting and higher survival rates, this projection may have converged (in the short term) on a $\lambda > 1$, and with more shooting and lower survival, the population would have declined, with $\lambda < 1$. After population growth reaches a stable $\lambda > 1$, total population size will increase geo-

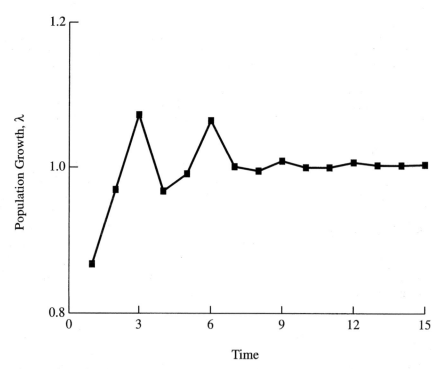

Figure 3.4   *Populus* plot of the population growth rate, λ vs t, projected in Table 3.4. Plotted values of λ give population growth, $\Sigma S_x(t)/\Sigma S_x(t-1)$, at each census. For this example, which begins arbitrarily with 24 newborn fawns, the population growth rate is initially variable, but then settles down to a constant value of $\lambda \approx 1$ as the projection proceeds.

metrically, as long as l_x and m_x remain constant. In the long term, increasing population size is likely to have density-dependent effects on survival and fertility.

Figure 3.5 is a *Populus* plot illustrating changes in population age structure for the same Table 3.4 projection. Like the growth rate, the proportion of individuals in the three illustrated age classes varies initially, and then steadies to a constant population composition called the stable age distribution. The age-specific rates of survival and fertility in the life table determine what proportion of this stable age distribution each age cohort comprises. Figure 3.6 illustrates a stable age distribution based on the McCullough (1979) l_x m_x schedule for George Reserve deer. *Populus* also produces a three-dimensional output graph, x vs $S_x/\Sigma S_x$ vs t, showing how the proportional composition of the population varies in time as it approaches stability, and a similar graph of x vs S_x vs t, showing temporal changes in the size of each cohort.

One practical consequence of a stable age distribution is that we can determine the l_x and m_x schedules in a single season. As long as age structure remains constant, S_x/S_0 gives the l_x for any age class, and F_x/S_x gives m_x. "Vertical" or "static" life tables constructed by this single-season technique are equivalent to "horizontal" or "cohort" life tables (constructed by following cohorts from birth to the disappearance of their last survivor) only if the age distribution is stable and constant.

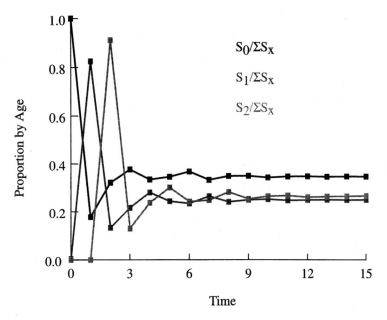

Figure 3.5   *Populus* plot of the proportional composition of the population projected in Table 3.4 by age. The black plot gives $S_0/\Sigma S_x$, the proportion of newborn fawns in the population, medium gray gives $S_1/\Sigma S_x$, the proportion of yearlings, and light gray plots $S_2/\Sigma S_x$, the proportion of two-year-old deer. Like the growth rates in Figure 3.4, these proportions are initially variable but converge to a Stable Age Distribution as the projection proceeds. With a different $l_x m_x$ schedule, proportional composition of this stable age distribution would change.

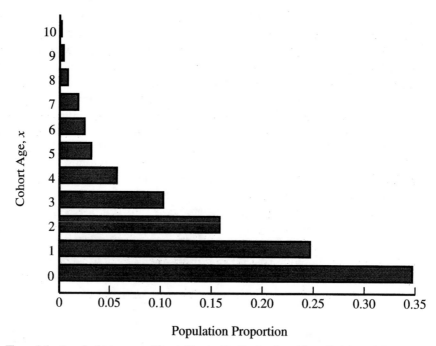

Figure 3.6   *Populus* histogram of the stable age distribution for white-tailed deer, *Odocoileus virginianus*, at Michigan's George Reserve estimated after 15 annual projection steps of the example in Table 3.4. Bars give the proportion of females in each age class. The l_x values for male deer are lower, and the proportion of bucks in older cohorts of a comparable stable age distribution for male deer would be reduced. Some texts plot male and female data to the left and right of a central *y*-axis, giving a figure that looks like a tree. Differences in the survival of males and females make the left and right branches asymmetrical.

3.4 AGE-STRUCTURED POPULATION GROWTH

If age-specific survival and fertility remain constant long enough for a population to reach a stable age distribution and constant growth rate, population size will change exponentially, and we can estimate this growth rate directly from demographic parameters. An important concept underlying these computations is the net reproductive rate, R_0, calculated (if the age distribution is stable) as

$$R_0 = \frac{\Sigma F_x}{S_0} \tag{3.3}$$

and defined as the average number of female progeny produced by each female in the original cohort, over her entire lifetime. The λ values calculated in the last section were discrete growth rates per single time step; in contrast, R_0 is a lifetime measure of reproduction. With $R_0 > 1$, a mother produces more than the single daughter required to compensate her death, and the population grows; with $R_0 < 1$ it shrinks, and with $R_0 = 1$ its size remains constant. When the interval of projection time steps is equal to the average generation length (which we will calculate below), then R_0 and λ are equivalent. Recalling that $F_x = S_x m_x$, we can substitute into the numerator of equation (3.3) and rearrange to show that the net reproductive rate is also a weighted average of m_x over all age classes,

$$R_0 = \frac{\Sigma S_x m_x}{S_0} = \Sigma \left(\frac{S_x}{S_0}\right) m_x \tag{3.4}$$

$$R_0 = \Sigma l_x m_x \tag{3.5}$$

Here the weighting term, l_x, is the probability of reaching age x, and m_x is the number of female progeny that a female produces if she does reach that age. Texts often present the net reproductive rate with equation (3.5), but (3.3) the sum of reproductive output produced by a cohort as it passes through all age classes, divided by initial newborn cohort size, seems more intuitive to many students (Lanciani 1987). When the age distribution remains stable, the numerator of equation (3.3) is also the sum of reproductive output produced concurrently by all of the age cohorts in a single census. Using equation (3.5) with the life table for George Reserve deer (Table 3.2), the net reproductive rate, R_0, is

$$R_0 = (1 \times 0) + (.71 \times .22) + (.46 \times .54) + (.3 \times .88) + (.17 \times .8)$$
$$+ (.07 + 1.15) + (.06 \times .46) + (.02 \times .81) + (.01 \times .92) + (.01 \times 0) = 1.01 \tag{3.6}$$

Thus, each doe replaced herself with an average of 1.01 female offspring, consistent with the observation that long-term population size was relatively constant.

Cohort-generation time, T_c is the average interval between the birth of mothers and their progeny, which is approximately

$$T_c \approx \frac{\Sigma x F_x}{\Sigma F_x} \tag{3.7}$$

The numerator is the sum of F_x, the number of progeny produced by each surviving cohort, multiplied by cohort age, x. With a stable age distribution it is also the sum of progeny produced by a single cohort as it passes through all age classes, again multiplied by age, x. Dividing by the sum of F_x values gives the maternal age at which progeny appear. An equivalent approximation of the cohort generation time that appears frequently in ecology texts is

$$T_c \approx \frac{\Sigma x l_x m_x}{\Sigma l_x m_x} \tag{3.8}$$

Using equation (3.8) with the life table for George Reserve deer (Table 3.2), we can estimate the cohort generation time as

$$T_c \approx \frac{(0 + .15 + .49 + .78 + .53 + .4 + .5 + .18 + .15 + .08 + 0 + 0)}{1.01} \approx 3.2 \tag{3.9}$$

In Figures 3.4 and 3.5, we showed that a population with a stable age distribution grows at a constant, density-independent rate. Knowing the cohort generation time, we can scale the discrete growth rate per generation, R_0, to a discrete population growth rate, λ, over any arbitrary time interval. In addition, because the instantaneous growth rate, r, is the natural logarithm of the finite λ defined over the same interval,

$$\lambda \approx \frac{R_0}{T_c} \quad \text{or} \quad r \approx \frac{\ln R_0}{T_c} \tag{3.10}$$

These approximations assume that T_c is quantified in the same units as λ and r. Their accuracy depends on the T_c estimate in the denominator. The estimate given by equations (3.7) or (3.8) is perfect for **semelparous** organisms that reproduce only once and then die, like members of the Pacific salmon genus *Onchorynchus*. For **iteroparous** organisms that contribute progeny to several successive cohorts, like the white-tailed deer, a T_c estimate is only accurate when population size is relatively stable, with $R_0 \approx 1$. When $R_0 \gg 1$ so that each female produces daughters over several breeding seasons, the first-born daughter begins contributing offspring to the growing population before later-born daughters. Equations (3.7) and (3.8) weight all daughters equally, giving an average T_c that will overestimate the real cohort-generation time, causing equation (3.10) to underestimate the population growth rate, r (May 1976).

Fortunately, there is a second, consistently accurate method of estimating the population growth rate from demographic data. It rests on the observation that populations with a constant r are changing size exponentially. Building on the discussion of exponential growth in Chapter 1, and assuming a stable age distribution, this means that the number of newborns (S_0) is changing exponentially over time.

$$S_0(t) = S_0(t-1)e^{r \cdot 1} = S_0(t-2)e^{r \cdot 2} = S_0(t-3)e^{r \cdot 3} = \ldots \tag{3.11}$$

Here $S_0(t)$ is the number of survivors of age zero (newborns) at time t, and $t-1$ is one time step prior to t. The exponents are r times 1, r times 2, and so on. Dividing both sides of the first equality by $e^{r \cdot 1}$, we have

$$\frac{S_0(t)}{e^{r \cdot 1}} = S_0(t-1) \tag{3.12}$$

Changing the sign of the exponent to put the $e^{r \cdot 1}$ term in the numerator and reversing the equation from left to right,

$$S_0(t-1) = S_0(t)e^{-r \cdot 1} \tag{3.13}$$

Likewise,

$$S_0(t-2) = S_0(t)e^{-r \cdot 2} \quad \text{and} \quad S_0(t-3) = S_0(t)e^{-r \cdot 3} \tag{3.14}$$

Setting these relations aside for a moment, the number of newborns at time t is the sum of progeny produced by one-year-old mothers, two-year-old mothers, three-year-old mothers, etc.

$$S_0(t)l_0 = S_0(t-1)l_1 m_1 + S_0(t-2)l_2 m_2 + S_0(t-3)l_3 m_3 + \ldots \tag{3.15}$$

Substituting equations (3.13) and (3.14) into (3.15), we have

$$S_0(t)l_0 = S_0(t)e^{-r \cdot 1}l_1 m_1 + S_0(t)e^{-r \cdot 2}l_2 m_2 + S_0(t)e^{-r \cdot 3}l_3 m_3 + \ldots \tag{3.16}$$

Dividing through to remove the $S_0(t)$ that appears in every term

$$l_0 = e^{-r\cdot 1}l_1 m_1 + e^{-r\cdot 2}l_2 m_2 + e^{-r\cdot 3}l_3 m_3 + \ldots \tag{3.17}$$

and because l_0, the probability of surviving from birth to birth, is 1.0,

$$\sum e^{-rx}l_x m_x = 1 \tag{3.18}$$

This is Euler's or Lotka's equation, named to honor the mathematician who derived it or the ecologist who first applied it in demographic analyses. We cannot solve equation (3.18) directly for r, but we can use it to check a guess or an estimate made with equation (3.10). The population growth rate for white-tailed deer at George Reserve estimated via equation (3.10) is

$$r \approx \frac{\ln(R_0)}{T_c} \approx \frac{\ln(1.01)}{3.2} \approx 0.003 \tag{3.19}$$

Substituting $r = 0.003$ and the $l_x\, m_x$ values of Table 3.2 into the Lotka-Euler equation gives

$$\sum e^{-rx}l_x m_x = 0 + .154 + .245 + .258 + .131 + .078$$
$$+ .082 + .025 + .019 + .009 + 0 + 0 = 1.001 \tag{3.20}$$

The close correspondence of this sum to 1.000 means that the r estimated in equation (3.19) is quite accurate. Although George Reserve deer are iteroparous, their population density was almost steady, with $R_0 = 1.01$, and we expect the cohort generation time given by equation (3.8) to be nearly perfect. Since the Lotka-Euler sum in equation (3.20) exceeds 1.0, the value of $r = 0.003$ is slightly too small. By entering formulas to calculate the Lotka-Euler sum in a spreadsheet program, it is easy to refine successive guesses, zeroing in on a very accurate value for r. The *Populus* program prints values for R_0, T_c, r (calculated by successive approximation using the Lotka-Euler equation (3.18) and (the estimated r) at the top of every output graph produced by its Age-Structured Population Growth module.

3.5 COLE'S PARADOX AND THE EVOLUTION OF LIFE HISTORIES

The life histories of organisms vary dramatically; semelparous Pacific salmon breed once and die, while iteroparous white-tailed deer breed repeatedly and may contribute progeny to nine successive cohorts. To explore this variation, it is useful to compare population-growth rates resulting from the hypothetical semelparous and iteroparous life tables in Table 3.5. The semelparous organisms on the left survive perfectly to reproductive age, produce 1 brood of 11 offspring, and die. The iteroparous organisms on the right survive perfectly through 9 breeding seasons, producing 10 offspring each season (one fewer than the semelparous brood size), and then die. For the semelparous case, the net reproductive rate is $R_0 = 11$, the cohort-generation time is $T_c = 1$, and the intrinsic growth rate estimated via equation (3.10) is $r = \ln R_0/T = \ln(11)/1 = 2.4$. Corresponding values for the iteroparous case are $R_0 \approx 90$, $T_c \approx \sum x l_x m_x/\sum l_x m_x \approx 450/90 \approx 5$, and $r \approx \ln R_0/T_c \approx \ln(90)/5 = 0.9$.

Table 3.5

HYPOTHETICAL LIFE TABLES ILLUSTRATING SEMELPAROUS AND ITEROPAROUS LIFE HISTORIES.

Semelparous				Iteroparous		
x	l_x	m_x		x	l_x	m_x
0	1.0	0		0	1.0	0
1	1.0	11		1	1.0	10
2	0	—		2	1.0	10
				3	1.0	10
				4	1.0	10
				5	1.0	10
				6	1.0	10
				7	1.0	10
				8	1.0	10
				9	1.0	10
				10	1.0	—

The cohort-generation-time estimates and population intrinsic growth rates arising from these examples are discussed in the text.

The first thing you should notice about these calculations is that the difference in r values doesn't make sense. The iteroparous organisms produce one fewer offspring at the same age as the semelparous organisms and subsequently reproduce eight more times. Why should r for the iteroparous case be so much lower? The answer is that it shouldn't. Both life tables give rapidly growing populations, and equation (3.10) badly overestimates T_c for the iteroparous case. The approximation of r becomes progressively worse for a growing population as the number of iteroparous broods and the time it takes to produce them increase. Checking the estimates of r for both cases by substitution into the Lotka-Euler equation, we find that the semelparous estimate is perfect but that population growth for the iteroparous case should be $r = 2.39$. The semelparous life history producing 11 offspring still results in a higher r than the iteroparous one with 9 intervals of 10 progeny each, but the quantitative difference is small.

Cole (1954) pointed out that both cases in Table 3.5 produce exactly the same growth rate if the iteroparous schedule is extended infinitely. In other words, by producing one extra offspring at age $x = 1$ and dying, a female achieves the same r as an immortal female producing 10 immortal daughters every year, forever. This is called "Cole's Paradox." The explanation for his counter-intuitive result lies in the peculiar survivorship schedule of Table 3.5, which assumes that juvenile and adult survival are equal (l_x values are 1.0). In fact, juveniles and adults seldom have the same probabilities of surviving to the next reproductive season; juveniles often have higher mortality in natural populations. This difference between age-specific mortality rates determines whether semelparity or iteroparity produces the highest population growth. Semelparity is favored when the probability that adults will survive between broods is lower than the chance of producing a sufficient number of additional offspring to offset the death of the parent. Iteroparity is favored when the probability that juveniles will survive to breeding age is low, but adult survival is high. Thus a high ratio of juvenile: adult mortality favors iteroparity, while a low ratio favors semelparity. For Pacific salmon, the cost of returning to the sea for a year, and then traveling back up the spawning rivers to reproduce again is extremely high, and the probability of success is low, favoring semelparity.

3.6 REPRODUCTIVE VALUE, V_X

We can gain additional perspective on the evolution of life histories by analyzing another demographic parameter that Fisher (1930, 1958) called V_x, the age-specific reproductive value. This is the expected number of future female progeny for a female of age x, relative to the expected future output of a newborn female, R_0. Figure 3.7 plots the reproductive value of George Reserve deer, which show a common pattern: Their reproductive value increases from birth to early adulthood and then declines. One reason causing V_x to rise from birth is that some newborns will not survive to reproductive age.

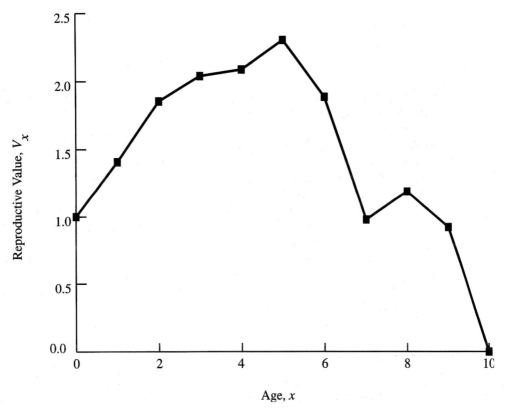

Figure 3.7 *Populus* plot of age-specific reproductive value, V_x, for white-tailed deer at Michigan's George Reserve. V_x is the age-specific lifetime expectation of daughters, relative to the expectation of a newborn female, R_0. In white-tailed deer, female ability to control high-quality breeding territory increases with age, causing V_x to increase well beyond the age of first reproduction.

The most common mathematical formulation for V_x,

$$V_x = \left(\frac{e^{rx}}{l_x}\right)\left(\sum_{y=x}^{\infty} e^{-ry} l_y m_y\right) \tag{3.21}$$

is a product with two parenthetic terms. The second term is part of the sum in the Lotka-Euler equation derived and explained in equations (3.11) to (3.18); this tallies the number of progeny produced by individuals of ages $y \geq x$. The first term, $e^{rx}/l_x = 1/e^{-rx} l_x$ divides this sum by the total number of females of age x in the current population. The current number of age x females is $e^{-rx} l_x$ because the population changes size exponentially in the interval between female birth and age x, and a fraction, l_x, of the original cohort will survive. Some properties of equation (3.21) are clearer if we substitute $l_x = S_x/S_0$, and $m_x = F_x/S_x$ and then simplify with algebraic rearrangements.

$$V_x = \frac{e^{rx}}{\dfrac{S_x}{S_0}} \sum_{y=x}^{\infty} e^{-ry} \frac{S_y}{S_0} \frac{F_y}{S_y} = \frac{e^{rx}}{S_x} \sum_{y=x}^{\infty} e^{-ry} F_y \tag{3.22}$$

$$V_x = \frac{\displaystyle\sum_{y=x}^{\infty} e^{r(x-y)} F_y}{S_x} \tag{3.23}$$

Equation (3.23) shows that the reproductive value is a weighted average of the number of female offspring produced at future ages (F_y) per female (S_x) (Lanciani 1987). In a growing population, $r > 0$, so the weighting term, $e^r(x-y)$, decreases with each succeeding age, y, and early offspring contribute more to reproductive value than their later siblings. Conversely, as the age, x, whose reproductive value we are evaluating increases from birth, the weight given to any particular age-specific reproductive contribution, F_y, increases. This is why reproductive value is lower for newborns than for those who are just reaching reproductive adulthood, even if there is no juvenile mortality. V_x finally falls to zero when a female's net reproduction is complete.

One theory of aging suggests that some deleterious genes have age-specific effects; those that compromise old individuals with low reproductive value will already have been passed to offspring when their negative effects occur. Natural selection against these genes will be ineffective, and senescence may result from an accumulation of such late-acting handicaps.

3.7 PROJECTION MATRICES

Prior to the widespread use of computers, the cumbersome projection techniques of Table 3.4 limited analyses of population age structure. To circumvent this problem, ecologists developed computational methods using the tools of matrix algebra. Most ecologists still use matrix projection because it simplifies complex demographic analyses, and the *Populus* software computes projections this way. Matrix projection requires survival and fertility parameters that allow population projections over a single time step. Let p_x be the probability of surviving from age x to age $x + 1$. For all of the surviving age classes, we can project forward from $S_x(t)$ the number of survivors in age class x at time t, as

$$S_{x+1}(t + 1) = p_x S_x(t) \tag{3.24}$$

Here $S_{x+1}(t + 1)$ is the number of survivors in age class $x + 1$ at time $t + 1$. A set of n such equations (where n is the maximum number of time steps survived by members of this population) would allow us to project abundance in all surviving age classes over a single time step.

Let f_x be the number of female progeny expected at time $t+1$ for every female of age-class x that is alive at time t. We can project the full cohort of new daughters expected at time $t + 1$ by adding the reproductive contributions of each surviving age class.

$$S_1(t + 1) = f_1 S_1(t) + f_2 S_2(t) + f_3 S_3(t) + \cdots + f_n S_n(t) \tag{3.25}$$

Notice that f_x is different from m_x (the expected number of female births per female during the interval from age $x - 1$ to age x) because it is affected by both fertility and survival. Not all of the m_x progeny born at time t will still be alive at time $t + 1$. The exact relations of f_x to m_x, and p_x to l_x differ for organisms with continuous or discrete reproduction, and for pre- and post-reproductive census timing. These details are covered below.

Equation (3.26) uses matrix notation to represent equation (3.25)and a set of equations

$$
\begin{pmatrix} S_1(t + 1) \\ S_2(t + 1) \\ S_3(t + 1) \\ \vdots \\ S_n(t + 1) \end{pmatrix} = \begin{pmatrix} f_1 & f_2 & f_3 & f & f_n \\ p_1 & 0 & 0 & \cdots & 0 \\ 0 & p_2 & 0 & \cdots & 0 \\ \vdots & \ddots & \ddots & \cdots & \vdots \\ 0 & 0 & 0 & p_{n-1} & 0 \end{pmatrix} \begin{pmatrix} S_1(t) \\ S_2(t) \\ S_3(t) \\ \vdots \\ S_n(t) \end{pmatrix} \tag{3.26}
$$

like equation (3.24). The leftmost and rightmost parenthetic terms in equation (3.26) are matrices with n rows and one column, representing the composition of the population at times $t + 1$ and t, respectively. Matrices with only one row or column are called "vectors." These vectors represent the population composition with S_x values, and the annual columns in Tables 3.1 and 3.4 give empirical and projected population-composition vectors for white-tailed deer. The square matrix with n rows and n columns is an "age-projection" matrix. Notice that the survival and fertility parameters of equations (3.24) and (3.25) comprise its non-zero elements. The matrix has many zero elements because individuals cannot grow younger or age more than one step at a time. Equation (3.26) states that multiplying the population–composition vector at time t by the projection matrix yields the population–composition vector at time $t + 1$ as its product.

A brief summary of matrix × vector multiplication will show how this projection process works. In order for them to be multiplied, the number of columns in a matrix must be the same as the number of rows in a vector. Their product is a new vector with the same dimension as the old one. Its first-row element is calculated by multiplying the first-row matrix elements by the corresponding column elements of the vector and adding the resulting products. This first step and the rest of the procedure for matrix × vector multiplication are summarized more clearly and easily in matrix notation,

$$
\begin{pmatrix} a_{11} & a_{12} \\ a_{21} & a_{22} \end{pmatrix} \begin{pmatrix} b_1 \\ b_2 \end{pmatrix} = \begin{pmatrix} a_{11}b_1 + a_{12}b_2 \\ a_{21}b_1 + a_{22}b_2 \end{pmatrix}
\tag{3.27}
$$

where the subscripts list first the row and then the column address of each element. For equation (3.26), the first-row element in the population-composition vector at time $t + 1$ is

$$
S_1(t + 1) = f_1 S_1(t) + f_2 S_2(t) + f_3 S_3(t) + \cdots + f_n S_n(t)
\tag{3.28}
$$

This is the number of progeny born to mothers of all age classes during the time interval from t to $t + 1$ that survive to time $t + 1$. Notice that this computation [equation (3.28)] for the first-row element of the product vector exactly reproduces the fertility projection of equation (3.25). The second-row element in the population-composition vector at time $t + 1$, $S_2(t + 1)$, is

$$
S_2(t + 1) = p_1 \times S_1(t) + 0 \times S_2(t) + 0 \times S_3(t) + \cdots + 0 \times S_n(t)
\tag{3.29}
$$

Since there is only one non-zero term in this summation, it simplifies to

$$
S_2(t + 1) = p_1 \times S_1(t)
\tag{3.30}
$$

In the same manner, the third-row element of the product vector is

$$
S_3(t + 1) = p_2 \times S_2(t)
\tag{3.31}
$$

and the computation of row 2 through row n elements of the product vector reproduce the survival projections made by n–1 copies of equation (3.24), one for each age class but the last. It is not necessary to project survival of the nth age class. We already know from the life table that there are no survivors.

All age-projection matrices have the same form. There are fertility parameters, $f_1 \cdots f_n$, in the top row indicating the reproductive contribution expected per female for each age class. The first column of row two holds p_1, the probability of surviving from age 1 to age 2. Single-step survival probabilities for older age classes, $p_2 \cdots p_{n-1}$, extend diagonally from p_1 toward the lower right. All other elements in the age-projection matrix are zeros. Leslie (1945, 1948) was among the first to use projection matrices of this form in demographic analyses, and we still call them "Leslie Matrices."

We can estimate values for the survival and fertility parameters, p_x and f_x, of the age-projection matrix from l_x and m_x; the exact method depends on whether the population has continuous or discrete, pulsed reproduction (Caswell 1989). For populations that breed in discrete pulses like white-tailed deer, all individuals within a cohort are roughly the same age, so we can estimate the survival and fertility parameters without averaging assumptions. The calculations depend on the timing of our census. Survival probabilities, p_x, are

$$p_x = \frac{l_x}{l_{x-1}} \quad \text{and} \quad p_x = \frac{l_{x+1}}{l_x} \tag{3.32}$$

for the post-reproductive and pre-reproductive censuses of Figures 3.1 and 3.3, respectively. The corresponding fertility parameter, f_x, for post-reproductive sampling is

$$f_x = p_x m_x \tag{3.33}$$

because females of age x must survive to the end of the projection interval (probability p_x) before producing m_x daughters *per capita*. With a pre-reproductive census, m_x daughters *per capita* have just been born, and their probability of surviving the projection interval is l_1, so

$$f_x = l_x m_x \tag{3.34}$$

For continuously reproducing populations, the length of time steps is arbitrary. The probability of surviving from age $x - 1$ to age x is l_x/l_{x-1}, but individuals in age class x can have any precise age between $x - 1$ and x. We therefore approximate p_x as the average l_x at the end of the age class, which is $(l_x + l_{x+1})/2$ over the average l_x at the beginning of the age class, $(l_{x-1} + l_x)/2$. Putting these together,

$$p_x \approx \frac{\left(\dfrac{l_x + l_{x+1}}{2}\right)}{\left(\dfrac{l_{x-1} + l_x}{2}\right)} \approx \left(\frac{l_x + l_{x+1}}{l_{x-1} + l_x}\right) \tag{3.35}$$

Fertility-parameter estimates, f_x, for a continuously reproducing population require a similar approximation. We assume that the average newborn must survive half of one projection interval to reach its first census, and the fertility of mothers between the ages of x and $x + 1$ is the weighted average of m_x and m_{x+1}. Then

$$f_x \approx l_{0.5}\left(\frac{m_x + p_x m_{x+1}}{2}\right) \tag{3.36}$$

where the parenthetic term is the average of offspring born at the beginning of the interval (m_x), and offspring born at the end of the interval (m_{x+1}), weighted by the probability that their mothers will survive (p_x); and $l_{0.5}$ is the probability that offspring will survive long enough to be counted,

$$l_{0.5} \approx \frac{l_0 + l_1}{2} \approx \frac{1 + l_1}{2} \tag{3.37}$$

3.8 STAGE-STRUCTURED POPULATIONS

Matrix projection is easily adapted to population analyses using categories other than age. For example, many perennial plants pass through a series of recognizable life-history stages, beginning as seeds, germinating to form vegetative stages that may grow for several years, and finally developing reproductive structures that produce new seeds. The precise duration of each stage can vary from plant to plant, depending on environmental conditions such as access to light, water, and soil nutrients. Seeds might germinate at their first opportunity or remain dormant in the soil for an extended period. Vegetative stages with ample resources may produce flowers and seeds from an

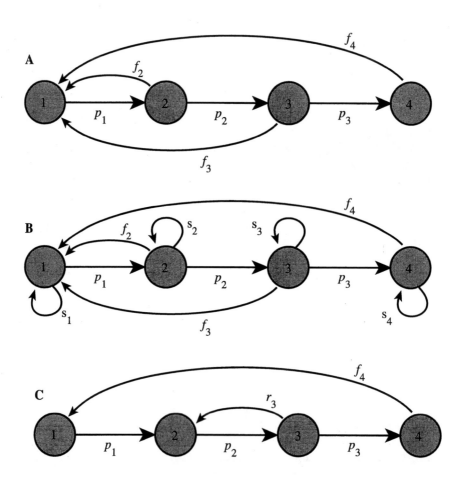

Figure 3.8   Life-cycle diagrams for stage-structured populations, redrawn after Caswell (1982, 1989). Case **A** represents a standard, age-structured life cycle. Case **B** is a size-structured cycle that includes probabilities of growing, or remaining in the same size class at each time step. Case **C** represents a cycle with both sexual and vegetative reproduction. Stage projection matrices corresponding to each example are presented in the text.

early age. In contrast, "century plants" of the desert southwest (*Agave kaibabensis*) grow as vegetative rosettes for many years before producing a single flower stalk and dying. It can be difficult or impossible to determine the age of such plants and much more convenient to base demographic analyses on the stages of their life history. Some plants reproduce both sexually and vegetatively, so the analysis of these life cycles can be complex.

Caswell (1982, 1989) introduced graphical techniques that make it easier to keep track of complex stage-structured life cycles. The first example in Figure 3.8 shows a Caswell life-cycle graph for an age-structured population like those of the previous sections but with four stages, represented by numbered circles. Arrows connect stage nodes i and j if individuals in stage i at time t can contribute individuals to node j at time $t + 1$ by surviving, growing, or reproducing. Survival and fertility parameters are defined just as they were in our previous examples, and the stage-projection matrix that corresponds to this life-cycle diagram is

$$\mathbf{A} = \begin{pmatrix} 0 & f_2 & f_3 & f_4 \\ p_1 & 0 & 0 & 0 \\ 0 & p_2 & 0 & 0 \\ 0 & 0 & p_3 & 0 \end{pmatrix} \tag{3.39}$$

Notice that it has the same form as the Leslie matrix of the previous section.

An important difference between stage- and age-structured populations is that the organisms may remain in one stage over several projection intervals. The second diagram in Figure 3.8 illustrates such cases. The coefficients p give the probabilities of moving from one stage to the next; s values give the probability of remaining in the same stage through the next time step; and f is the fertility coefficient, as before. The matrix that we use to project the changing composition of this population over time must include the probabilities of advancing to a new stage and remaining in the same stage at each time step. Probabilities of advancing, p, run diagonally across the age projection matrix in positions where the row i and column j addresses are $i = j + 1$. Probabilities of remaining in place at the next census, s, go diagonally across the projection matrix in positions where $i = j$. The stage-projection matrix corresponding to life cycle B of Figure 3.8 is

$$\mathbf{B} = \begin{pmatrix} s_1 & f_2 & f_3 & f_4 \\ p_1 & s_2 & 0 & 0 \\ 0 & p_2 & s_3 & 0 \\ 0 & 0 & p_3 & s_4 \end{pmatrix} \tag{3.40}$$

Life-cycle graphs can illustrate natural histories that are even more complex. All of the projection matrices we have studied to this point confine reproductive parameters to the top row; new individuals are always in stage 1 when first counted. This is not necessarily the case in stage-structured populations. The third case in Figure 3.8 represents a life cycle where individuals produce stage-1 progeny if they reach stage 4 but also produce stage-2 progeny if they reach stage 3. These two kinds of progeny could

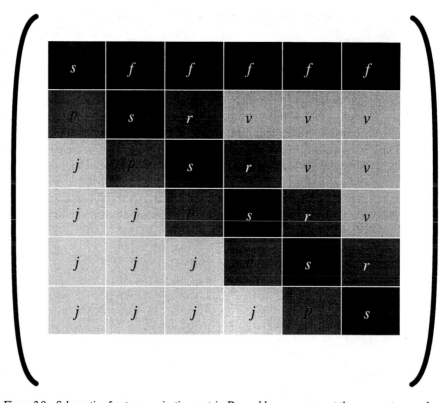

Figure 3.9   Schematic of a stage-projection matrix. Row addresses represent the source stage, and column addresses represent the destination stage for the various transitions. Dark elements labeled *s* on the main diagonal, where row and column addresses are equal, are probabilities of stasis, or remaining in the same stage at the next time step. Gray elements labeled *p* are probabilities of advancing one stage, like the survivorship elements of a Leslie matrix. Light elements, *j*, are probabilities of jumping two or more stages per time step. These elements always carry zero probabilities in an age-projection matrix, and will often be zeros in stage matrices as well. Gray elements labeled *r* give probabilities of regression by one stage. Plant demographers may find non-zero probabilities here for individuals that suffer significant tissue loss to herbivores. Elements labeled *r* and *v* might also represent vegetative reproduction contributing to stages later than stage 1. Top-row elements, *f*, give fertility, reproductive contributions to stage 1, like the fertility elements of the Leslie matrix. Stages are not necessarily numbered in the sequence of their appearance in the life cycle.

be seeds and vegetative tillers, respectively. Now the projection matrix will have reproductive parameters below its top line.

$$\mathbf{C} = \begin{pmatrix} 0 & 0 & 0 & f_4 \\ p_1 & 0 & r_3 & 0 \\ 0 & p_2 & 0 & 0 \\ 0 & 0 & p_3 & 0 \end{pmatrix} \tag{3.41}$$

It is also possible for a plant to regress to some earlier stage. For example, a large vegetative plant that suffers heavy herbivory and loses much of its biomass may regress to a small vegetative stage. Figure 3.9 gives a schematic representation of the stage-projection matrix summarizing additional possibilities.

These examples of age- and stage-structured population growth demonstrate that the tools of matrix projection are quite flexible, facilitating study of both simple and complex life cycles. They also allow conclusions to be drawn about the dynamics of structured populations, directly from properties of the projection matrix. Advanced students who wish to explore these techniques further will find valuable guidance in Caswell (1989) and Carey (1993).

REFERENCES

Carey, J. R., *Applied Demography for Biologists with Special Emphasis on Insects*. Oxford: Oxford Universtiy Press, 1993.

Case, T. J., *An Illustrated Guide to Theoretical Ecology*. New York: Oxford University Press, 2000.

Caswell, H, "Stable Population Structure and Reproductive Value for Populations with Complex Life Cycles." *Ecology*, 1982, 63:1223–1231.

Caswell, H, *Matrix Population Models*. Sunderland MA: Sinauer Associates, 1989.

Cole, L. C., "Population Consequences of Life History Phenomena." *Quart. Rev. Biol.*, 1954, 29:103–137.

Emlen, J. M., *Population Biology: The Coevolution of Population Dynamics and Behavior*. New York: Macmillan Publishing Company, 1984.

Fisher, R. A., *The Genetical Theory of Natural Selection*. Dover Publications Inc., 1958.

Jenkins, S. H. "Use and Abuse of Demographic Models of Population Growth." *Ecological Society of America Bulletin*, 1988, 69:201–207.

Lanciani, C. A. "Teaching Quantitative Concepts of Population Ecology in General Biology Courses." *Ecological Society of America Bulletin*, 1987, 68:492–495.

Leslie, P. H. "On the Use of Matrices in Certain Population Mathematics." *Biometrika*, 1945, 33:183–212.

Leslie, P. H. "Further Notes on the Use of Matrices in Population Mathematics." *Biometrika*, 948, 35:213–245.

Manly, B. F. J., *Stage-Structured Populations: Sampling, Analysis and Simulation*. London: Chapman and Hall, 1990.

May, R. M., "Estimating *r*: A Pedagogical Note." *American Naturalist*, 1976, 110:496–499.

McCullough, D. R., The George Reserve Deer Herd: Population Ecology of a *K*-Selected Species. Ann Arbor: The University of Michigan Press, 1979.

Reed, M. J., P. D. Doerr and J. R. Walters, "Minimum Viable Population Size of the Red-Cockaded Woodpecker." *J. Wldl. Manage*. 1988, 53(3):385–391.

Stearns, S. C., *The Evolution of Life Histories*. Oxford: Oxford University Press, 1992.

Wilson, E. O. and W. H. Bossert, *A Primer of Population Biology*. Sunderland MA: Sinauer Associates, 1971.

PROBLEMS AND EXERCISES

LIFE TABLES

1.   Select Age-Structured Population Growth from the *Populus* models menu, and enter the following $l_x m_x$ schedule in the input table.

x	l_x	m_x
0	1.00	0
1	0.50	1
2	0.25	3
3	0	0

Set the runtime to $t = 10$ projection steps so that the simulation has sufficient time to approach a stable age distribution and constant growth rate.

(a) Run the model with these parameter values, and make a list of all the different output views provided by the *Populus* program, explaining the significance of each view.

(b) Estimate the equilibrial value of λ from the computer screen and the equilibrial proportion of individuals in each age class, $S_x/\Sigma S_x$.

(c) Why do the values of λ and $S_x/\Sigma S_x$ oscillate initially? Is there any set of initial S_x values that would prevent this oscillation?

(d) When the age distribution stabilizes, is the population growing, declining or remaining constant?

(e) What is the Net Reproductive Rate, R_0, for the default example?

(f) Now, change the m_x so that $m_1 = 0$ and $m_2 = 4$; then run the model again. Does the age distribution stabilize? Explain this result. Can you think of any natural populations that might follow this pattern?

(g) What is the Net Reproductive Rate, R_0, for the revised m_x schedule in (f) above?

2.   Devise your own life tables to produce (a) a growing population, (b) a declining population, and (c) a stable population. Run all three simulations for 10 time steps, and describe the graph of ΣS_x vs t, the time trajectory of total population size, for each run.

3.   There are two ways to estimate a population intrinsic-growth rate, r, using life-table information. One method uses the Lotka-Euler equation (3.18) yielding an accurate r-value by successive approximation. The second method is easier but may be less accurate; this involves estimation of r as $r \approx \ln R_0/T_c$, where T_c is the cohort generation time, approximated by equations (3.7) or (3.8). The accuracy of this approximation varies among life tables. Run a single l_x schedule with three different m_x schedules, all listed in the table below. Record the net reproductive rate, R_0, the cohort generation time, T_c, the population intrinsic-growth rate, r, yielded by successive approximation with the Lotka-Euler equation, and the r-value approximated as $r \approx \ln R_0/T_c$.

x	l_x	m_x	m_x	m_x
0	1.0	0	0	0
1	0.5	8	4	4
2	0.4	0	2	0
3	0.3	0	2	0
4	0.2	0	2	0
5	0.1	0	2	20
6	0	—	—	—

For which of the m_x schedules is the estimate using $r \approx \ln R_0/T_c$ best? Which is worst? Is the accuracy of r estimates affected by the net reproductive rate in these examples? What general conditions cause equations (3.7) or (3.8) to give good and bad estimates of the cohort generation time?

LIFE HISTORIES

4. How do the number of offspring per breeding attempt (litter or clutch size) and the lifetime number of breeding attempts affect the population intrinsic growth rate? Look at the following l_x m_x schedules. Make and explain a prediction about the relative values of r that you expect for each schedule, and then run the model with these values to check your expectation.

x	l_x	m_x	x	l_x	m_x	x	l_x	m_x
0	1.00	0	0	1.00	0	0	1.00	0
1	0.50	3	1	0.50	4	1	0.50	3
2	0.25	0	2	0.25	0	2	0.25	3
3	0	0	3	0	0	3	0	0
	$r =$			$r =$			$r =$	

5. The age at which reproduction takes place can have a major effect on the growth of an age-structured population. The columns below present a single l_x schedule, and four different m_x schedules. Note that each has an equivalent R_0; the only difference is in the timing of reproduction, where reproductive output, m_x, is set to zero at varying ages. Make and explain a prediction about the relative population-growth rates of the four schedule combinations. Then run the following cases, record the value of r to test your expectations, and summarize your findings.

x	l_x	m_x	m_x	m_x	m_x
0	1	0	0	0	0
1	0.5	3	0	3	3
2	0.5	3	3	0	3
3	0.5	3	3	3	0
4	0.5	0	3	3	3
5	0	—	—	—	—
		$r =$	$r =$	$r =$	$r =$

6. "Cole's paradox" suggests that the growth rate of immortal organisms that have n progeny each year is equivalent to the growth rate of organisms that have $n + 1$ offspring and die after their first reproductive season. Cole invoked no differences between juvenile and adult mortality, which help resolve the paradox. Use the values listed below to make two different comparisons of iteroparous and semelparous life histories. In both cases, you are contrasting a population that produces 6 offspring and survives to reproduce again with one that produces 7 offspring at the cost of early mortality. On the left and right, this contrast is made between two populations with markedly different l_x schedules.

x	l_x	m_x	m_x		x	l_x	m_x	m_x
0	1	0	0		0	1	0	0
1	0.1	6	7		1	0.9	6	7
2	0.1	6	0		2	0.45	6	0
3	0.1	6	0		3	0.1	6	0
4	0	—	—		4	0	—	—
		$r =$	$r =$				$r =$	$r =$

Find all four values of r and use your own words to draw and explain general conclusions about the conditions that favor iteroparity or semelparity based on juvenile (before first reproduction) and adult mortality.

REPRODUCTIVE VALUE

7. The reproductive value of mammals typically increases from birth to the age of first reproduction. Most texts attribute this increase to infant mortality (cf. Wilson and Bossert 1971, p. 122), i.e. they assert that the average reproductive value of newborns is lower than that of young adults because some don't survive to reproduce. Would the pattern of reproductive value be different if survival to the age of first reproduction was perfect? Formulate several life tables to explore the causes of this typical mammalian pattern. Determine the age-specific reproductive values, either with a pencil or with *Populus* simulations, and explain your results.

PROJECTION MATRICES

8. Make a Leslie age-projection matrix based on the default l_x m_x schedule given in question 1, above. Then make a population–composition vector for an initial population comprised of only a single, newborn individual. Multiply the matrix and vector with pencil and paper to produce a new population–composition vector for time $t + 1$. Explain the biological significance of each step in the multiplication process, and each element in the projection matrix and vectors.

9.   Can you propose a Leslie matrix that will cause population age structure to oscillate indefinitely, never settling into a stable age distribution (like the case in problem 1f)? Can you propose a general rule that will produce additional matrices with this same behavior?

PROBLEMS WITH EMPIRICAL DATA

10.   Enter the life table for white-tailed deer at the George Reserve in a spreadsheet, and write a spreadsheet formula to make the summation for the Lotka-Euler equation, $\Sigma e^{-rx} l_x m_x$, automatically. Use your spreadsheet to check the value of r given in equation (3.19). Then try different values of r to determine whether equation (3.19) was rounded upward or downward to its precision of 3 decimal places.

11.   The projection in Table 3.4 assumes a post-reproductive census. How would this projection differ if the data were assumed to result from pre-reproductive censuses? Would this raise or lower our estimate of the population growth rate?

12.   Reed *et al.* (1988) presented life-table data for red-cockaded woodpeckers (*Picoides borealis*) in the southeastern U.S. Based on long-term observation of marked individuals, they present the following $l_x\, m_x$ schedule for females:

x	l_x	m_x
0	1.000	0.000
1	0.324	0.370
2	0.231	0.795
3	0.181	0.900
4	0.127	0.895
5	0.089	0.930
6	0.062	0.930
7	0.043	0.930
8	0.030	0.930
9	0.021	0.930
10	0.015	0.930
11	0	—

Enter these life-table data on the *Populus* input screen, and simulate the dynamics. Is this schedule based on pre- or post-reproductive censuses? Is this red-cockaded woodpecker population growing or declining? If conservation efforts could be targeted to enhance the survival and reproduction of particular age classes, which age class(es) should we target, and how much change is necessary? Use the *Populus* software to run a sensitivity analysis, replacing the observed l_x and m_x values with hypothetical test numbers that will help answer these questions.

PROBLEMS FOR ADVANCED STUDENTS

13. Suppose you were to study a species whose birth and death occur in pulses like those of white-tailed deer. What would happen if you set a time interval different from the interval of birth and death pulses to divide the life cycle into cohort groups? Consider and discuss the implications of this mismatched timing.

14. Here is a stage-projection matrix redrawn after Caswell (1982, 1989):

$$\begin{pmatrix} 0 & 0 & 0 & 0 & f_5 \\ p_1 & 0 & 0 & 0 & 0 \\ j_1 & 0 & 0 & r_4 & 0 \\ 0 & j_2 & p_3 & 0 & 0 \\ 0 & 0 & 0 & p_4 & 0 \end{pmatrix}$$

Can you produce a corresponding life-cycle graph like those in Figure 3.8 and explain the biological processes illustrated by this example? What are the constraints on possible values of p_1 and p_3?

CHAPTER 4
Lotka-Volterra Competition

Organisms interfere with one another in many ways. Tall trees that form a closed forest canopy block light, preventing or slowing the growth of seedlings in the shaded understory. Yeast cells growing in grape juice meet their energetic requirements by metabolizing simple sugars, converting them to ethanol. As the yeast population grows, sugar concentration declines until both population growth and fermentation stop. Finally, starved and encysted yeast cells settle from suspension, and we drink the wine. In both examples, organisms are restricting neighbors' use of a resource, some requirement that is necessary for population growth, either by blocking access or by consuming that resource themselves. We refer to this negative interaction between organisms as "competition."

The logistic growth models of Chapter 2 illustrated effects of **intraspecific** (within-species) competition, incorporating a negative feedback of population size on population growth. As a population with logistic growth increases in density, individuals suffer more competition from other members of the population, and their *per capita* birth and survival rates decline. **Interspecific** (between-species) competition is a directly analogous process. All plants need water, light, and nutrients, and the consumption of those essentials by a neighbor can limit individual growth, even if the neighbor is a member of some other species. In this chapter, we develop a model of competition based on interactions within and between two species.

4.1 EMPIRICAL EXAMPLES

Competition implies an inverse relation between population density and *per capita* population growth, like the examples shown in Figure 2.2. At the end of Chapter 2, we learned that environmental variability and the oscillation caused by lags or discrete life histories make it difficult to demonstrate density-dependent feedback by direct observation of natural population dynamics. For this reason, the best evidence of competition comes from experimental manipulation, and there is a long tradition of competition experiments in ecology (*cf*. Tansley 1917; Gause 1934; Park 1948, 1954; Connell 1961a, b). Figure 4.1 illustrates an experiment by Tilman *et al.* (1981) in which the diatoms *Asterionella formosa* and *Synedra ulna* competed in laboratory culture. The mixed, temperature-controlled cultures were semicontinuous, with a fraction of each culture decanted at constant intervals and replaced with fresh medium. The single limiting nutrient in that medium was silicate, an essential component of the diatom cell wall. Cultures received an inoculation of one or both species. In the absence of interspecific competitors, both *Synedra* and *Asterionella* grew logistically, increasing to a constant, equilibrial density. *Per capita* population growth declined with increasing diatom abundance, giving a clear example of intraspecific competition. In combined

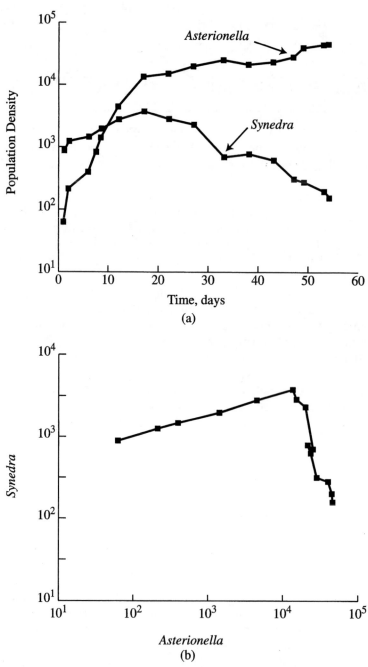

Figure 4.1   Laboratory competition experiment between the diatoms *Asterionella formosa* and *Synedra ulna*, limited by the concentration of silicate available in semicontinuous culture. A constant fraction of the growth medium was replaced each day. Initial density of *Synedra* (in cells per ml) was higher than *Asterionella*. (a) Both populations grew initially, but after *Asterionella* became abundant, *Synedra* gradually disappeared from the culture (*N* vs *t* plot redrawn after Tilman, Mattson and Langer, 1981). (b) A corresponding plot of the same data with *Synedra* densities on the *y*-axis, and *Asterionella* densities on the *x*-axis.

culture, both species increased initially, but after *Asterionella* became abundant, *Synedra* started to decline and was ultimately driven to extinction. Under these experimental conditions, *Asterionella* was superior, competitively excluding *Synedra*.

Similar experiments are possible in natural populations. Brown *et al.* (1986) studied plants and the community of coexisting ant and rodent species that eat their seeds in the Sonoran and Chihuahuan Deserts of Arizona. They fenced experimental plots and established long-term treatments including partial and complete rodent removal, ant removal, and seed augmentation, with unmanipulated control plots. By tracking density compensations or changes in the abundance of unmanipulated species in each experimental treatment, Brown and his students were able to demonstrate direct competition between large and small rodents, ants and rodents, and small- and large-seeded plants. There were other results of the study, which every ecology student should read, but this brief summary shows that competition can have a dramatic effect on community composition, and it does not always result in competitive exclusion. Four genera of granivorous rodents and five genera of granivorous ants coexisted in these desert communities.

Thus, competitive interaction has two possible outcomes: coexistence or exclusion of the inferior competitor. One objective for our competition model will be to predict what conditions lead to one or the other outcome.

4.2 LOTKA-VOLTERRA COMPETITION

The competition model that we introduce here is a simple extension of the continuous logistic growth model from Chapter 2. In that model, we codified mutual interference with a negative feedback term, $(K - N)/K$, representing the fraction of environmental carrying capacity unused by a population of size N, measured in units of individuals. The effect of this term caused the continuous logistic model,

$$\frac{dN}{dt} = rN\left(\frac{K - N}{K}\right) \tag{4.1}$$

to regulate population size at a stable equilibrial density, K. By using subscripts to distinguish two competing species, we can add both intra- and interspecific effects to this feedback term, $(K_1 - (N_1 + N_2))/K_1$. Now N_1, the density of species 1, and N_2, the density of species 2, are both subtracted from K_1, the carrying capacity for species 1. Since 1 and 2 are different species with different biological properties, they are unlikely to make exactly equivalent demands on the habitat, and it is customary to add a scaling factor, α, to account for these differences, $(K_1 - (N_1 + \alpha N_2))/K_1$. If species 1 is a bird that eats mostly seeds and a few insects while its competitor, species 2, eats mostly insects and a few seeds, then each interspecific competitor will cause less interference than an intraspecific competitor, and $\alpha < 1$. This scaling factor, α, is called a **competition coefficient**; it quantifies the *per capita* reduction in population size of species 1 caused by species 2.

We can use this revised feedback term to build an analog of the continuous logistic equation.

$$\frac{\mathrm{d}N_1}{\mathrm{d}t} = r_1 N_1 \left(\frac{K_1 - (N_1 + \alpha N_2)}{K_1} \right) \tag{4.2}$$

This equation specifies the population growth of species 1, with effects of both intraspecific competition and interspecific competition from species 2. Rearrangement of equation (4.2),

$$\frac{\mathrm{d}N_1}{N_1 \mathrm{d}t} = r_1 - \frac{r_1}{K_1}(N_1 + \alpha N_2) \tag{4.3}$$

shows that the *per capita* growth of species 1 is a linearly declining function of both species-1 density (N_1) and species-2 density (N_2). The intensities of intraspecific and interspecific feedback differ by the scaling factor, α.

Three terms are multiplied together on the righthand side of equation (4.2). The first, r_1, is the intrinsic growth rate of species 1; N_1 is the number of species-1 individuals present, and the parenthetic feedback term slows population growth as the interference of intra- and interspecific competitors increases. If $\alpha N_2 \geq N_1$, the value of the feedback term will be zero or negative, even for very small values of N_1. This means that interspecific competition from species 2 can limit or eliminate species 1, even without intraspecific feedback. In Figure 4.1, *Asterionella* probably has this effect on *Synedra*. After about day 20, $N_1 + \alpha N_2 > K_1$, and the *Synedra* feedback term becomes negative.

Since competition is an interaction between two species, there is an analogous equation giving the dynamics of species 2.

$$\frac{\mathrm{d}N_2}{\mathrm{d}t} = r_2 N_2 \left(\frac{K_2 - (N_2 + \beta N_1)}{K_2} \right) \tag{4.4}$$

There is no reason to expect that the reciprocal interference of two species will be precisely equal, so we define β, a second competition coefficient quantifying the *per capita* displacement of species 2 by species 1. Together, equations (4.2) and (4.4) are the Lotka-Volterra competition equations, named in honor of two ecologists who proposed them independently (Lotka 1925, 1932; Volterra 1926).

4.3 DYNAMICS OF LOTKA-VOLTERRA COMPETITION

To project the future dynamics of a single population with exponential or logistic growth, we derived a definite integral specifying population size as a function of time. Unfortunately, paired differential equations like (4.2) and (4.4) have no such solution. With both N_1 and N_2 appearing in the $\mathrm{d}N_1/\mathrm{d}t$ equation and the $\mathrm{d}N_2/\mathrm{d}t$ equation, the dynamics are intertwined. Any population size reached by species 1 affects the growth rate of species 2, and vice versa. There are several routes around this problem. One approach is numerical integration, using a computer to sum instantaneous changes in population size. *Populus* uses this technique to produce graphical pictures of Lotka-Volterra competition. Figure 4.2 illustrates a *Populus* simulation based on the default

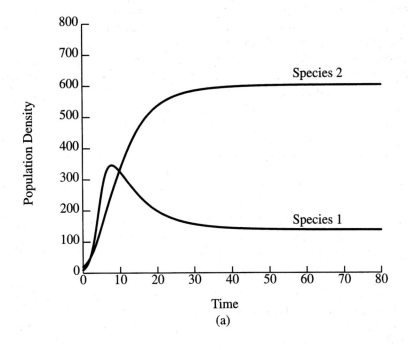

(a)

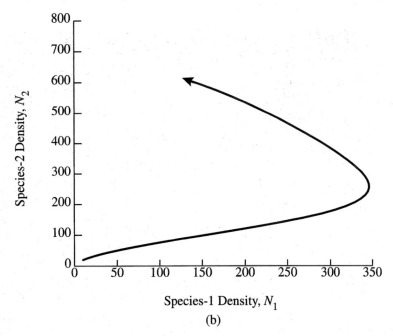

(b)

Figure 4.2   (a) *Populus* time trajectory of Lotka-Volterra competition based on the default parameter values of Box 4.1. This example shows a case resulting in stable coexistence. Species 1 achieves higher initial growth, but falls back with increasing interspecific competition. Ultimately, both species reach stable, equilibrial densities. (b) Corresponding graph of the same simulation on an N_2 vs N_1 phase plane.

parameter values of Box 4.1. The upper graph is a familiar time trajectory (N_1 and N_2 plotted against t). The lower graph is a *phase plane*, representing the same data. A phase plane is a standard mathematical tool illustrating the dynamics of paired differential equations, plotting one variable on each axis. In this case, N_2 is plotted on the y-axis, with N_1 on the x-axis. The plot moves to the right as N_1 increases and then reverses as N_1 falls. At the same time, N_2 increases steadily to an equilibrium density near $N_2 = 600$. The lower graph in Figure 4.1 shows a corresponding phase-plane plot of the interaction between *Asterionella* and *Synedra*.

BOX 4.1

The *Populus* input window for Lotka-Volterra Competition, with parameter values and option settings that produce the simulation of stable two-species coexistence illustrated in Figure 4.2a. Selecting the N_2 vs N_1 plot-type option gives the phase-plane output graph shown in Figure 4.8.

This simulation requires initial population sizes, intrinsic growth rates, carrying capacities, and competition coefficients for each species. Each of the values can be toggled up or down; the coupled output shows changes immediately.

The Lotka-Volterra Competition module offers 2 outputs, a time trajectory of population sizes and a phase-plane graph, with N_2 on the y-axis and N_1 on the x-axis. The phase plane shows both isoclines and changing abundances.

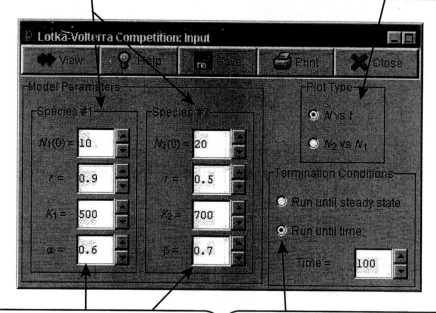

The competition coefficients α and β quantify the *per capita* reduction in equilibrial population size caused by interspecific competitors.

The simulation can be set to terminate after a fixed run time or allowed to run until N_1 and N_2 are no longer changing

4.4 ISOCLINE ANALYSES

A second method for studying the dynamics of Lotka-Volterra competition is to characterize the equilibria of the equations. At an equilibrium of equations (4.2) and (4.4), both

$$\frac{dN_1}{dt} = 0 \quad \text{and} \quad \frac{dN_2}{dt} = 0 \tag{4.5}$$

When these two conditions are both true, the dynamics of competition are at rest. We can also characterize the stability of these equilibria. The densities of competing species move toward stable equilibria and away from unstable ones. Stable equilibria identify outcomes of competitive interaction.

First, we will examine the conditions that lead to equilibria for species 1. Setting equation (4.2) equal to zero, we have

$$\frac{dN_1}{dt} = 0 = r_1 N_1 \left(\frac{K_1 - (N_1 + \alpha N_2)}{K_1} \right) \tag{4.6}$$

Since three terms are multiplied together on the righthand side of this equation, there are three solutions that fulfill the equality. Two occur when $r = 0$ or $N_1 = 0$. These are trivial cases of a population without members or growth potential. A more interesting case is the one where the value of the parenthetic feedback expression is equal to zero:

$$0 = \left(\frac{K_1 - (N_1 + \alpha N_2)}{K_1} \right) \tag{4.7}$$

Algebraic rearrangements

$$0 = K_1 - N_1 - \alpha N_2 \tag{4.8}$$

$$N_2 = \frac{K_1}{\alpha} - \frac{1}{\alpha} N_1 \tag{4.9}$$

show that this is the equation for a straight line (equation 4.9). Plotting N_1 on the x-axis and N_2 on the y-axis of Figure 4.3a, the y-intercept is K_1/α, the x-intercept is K_1, and the slope is $-1/\alpha$. This line, called the zero-net-growth isocline for species 1, is the set of all combinations of N_1 and N_2 for which $dN_1/dt = 0$. This means that the species-1 *per capita* birth and death rates are equal. If competitor 2 is absent, the equilibrial density of competitor 1 is K_1. With each species-2 individual added to the habitat, the equilibrial density of species 1 falls by the amount α.

Points below and to the left of the species-1 isocline in Figure 4.3a represent combined densities of both species that are too small to prevent population growth of species 1. In this case, $N_1 + \alpha N_2 < K_1$, so both the parenthetic feedback term of equation (4.6) and dN_1/dt will be positive. Species 1 can increase in density. Conversely, for points above and to the right of the species-1 isocline, dN_1/dt is negative, and species-1 density will decrease.

Our isocline analysis has illustrated only the dynamics of competitor 1. To understand the interaction fully, we need a corresponding analysis for competitor 2. Setting

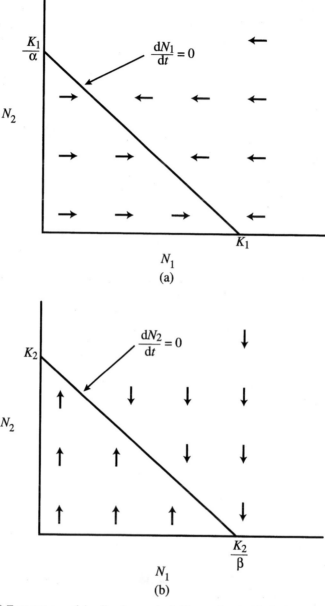

Figure 4.3 (a) Zero-net-growth isocline for species 1. From points inside the isocline, $dN_1/dt > 0$ because the combination of intraspecific interference from species 1 and interspecific interference from species 2 is not sufficient to prevent growth by species 1. Growth vectors in this zone point to the right, parallel to the N_1 axis, because N_1 is increasing. From points outside the isocline, $dN_1/dt < 0$, and species-1 density falls toward the isocline. Growth vectors point to the left because N_1 growth is negative in this zone. (b) Zero-net-growth isocline for species 2. From points inside the isocline, $dN_2/dt > 0$, because the combination of intraspecific interference from species 2 and interspecific interference from species 1 is not sufficient to prevent growth by species 2. Growth vectors in this zone point up, parallel to the N_2 axis, because N_2 is increasing. From points outside the isocline, $dN_2/dt < 0$ and species-2 density falls toward the isocline. Growth vectors point down because N_2 growth is negative in this zone.

equation (4.4) equal to zero and repeating the manipulation of equations (4.6) through (4.9), we have

$$\frac{dN_2}{dt} = 0 = r_2 N_2 \left(\frac{K_2 - N_2 - \beta N_1}{K_2} \right) \tag{4.10}$$

$$0 = \left(\frac{K_2 - N_2 - \beta N_1}{K_2} \right) \tag{4.11}$$

$$0 = K_2 - N_2 - \beta N_1 \tag{4.12}$$

$$N_2 = K_2 - \beta N_1 \tag{4.13}$$

Again, this is the equation for a straight line. Plotting N_2 on the y-axis and N_1 on the x-axis (Figure 4.3b) gives a y-intercept of K_2, an x-intercept of K_2/β, and a slope of $-\beta$. This is the isocline for species 2, giving the set of all combinations of N_1 and N_2 for which $dN_2/dt = 0$. This means that the species-2 *per capita* birth and death rates are equal. If competitor 1 is absent, the equilibrial density of competitor 2 is K_2. With each species-1 individual added to the habitat, the equilibrial density of species 2 falls by the amount β. From points below and to the left of the isocline (shown in Figure 4.3b) species 2 increases, and from points above and to the right, it decreases.

4.5   COEXISTENCE OR DISPLACEMENT?

By superimposing species-1 and species-2 isoclines on the same phase-plane graph, we can observe the equilibria of both species simultaneously. The two isoclines have four possible arrangements, shown in Figures 4.4–4.7. In Case I (Figure 4.4) the isoclines do not cross; the species-1 isocline lies entirely above and to the right of the species-2 isocline. This means that we could begin at any initial densities with species 1 and species 2 both present, and this arrangement would lead to a stable equilibrium at $N_1 = K_1$, and $N_2 = 0$. Species 1 is the superior competitor, and the result of their interaction is competitive exclusion of species 2.

The intercepts of Figure 4.4 show why species 1 wins this interaction. First, compare the x-intercepts (where $N_2 = 0$). K_1 is the carrying capacity, the number of species-1 individuals required to prevent further increase of species 1. K_2/β is the number of species-1 individuals required to prevent the increase of species 2 (at the limit where $N_2 = 0$). Since $K_2/\beta < K_1$, the interspecific effect of species 1 on species 2 is stronger than the intraspecific effect of species 1 on itself. This means that species 1 can increase to $N_1 = (K_2/\beta)$, and then continue increasing to the higher density of $N_1 = K_1$ before experiencing self-limitation.

This situation is reversed for the y-intercepts of Figure 4.4 (where $N_1 = 0$). K_2 is the number of species-2 individuals required to prevent further increase of species 2. K_1/α is the number of species-2 individuals required to prevent the increase of species 1 (at the limit where $N_1 = 0$). Since $K_2 < K_1/\alpha$, the inhibition of species 1 by species 2 is weaker than the self-inhibition of species 2. This means that species 2 will reach its own carrying capacity before reaching the density necessary to limit species 1.

Case I

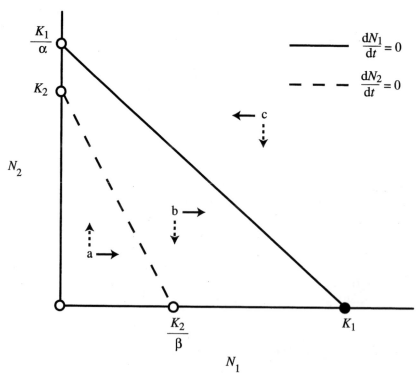

Figure 4.4   **Case I.** In this arrangement, the isocline of species 1 lies entirely above and to the right of the species-2 isocline. If N_1 and N_2 densities are those given by point a, then the combined numbers are too small to limit the growth of either species and both will increase in abundance. Point b is above and to the right of the species-2 isocline but below the species-1 isocline. This means that $dN_2/dt < 0$ but $dN_1/dt > 0$ and the increase of species 1 can continue. For all points on the species-2 isocline $dN_2/dt = 0$; again, species 1 can still increase. Both species decline from the densities indicated by point c. The interaction of these two competitors results in elimination of species 2, with species 1 increasing to its equilibrial density K_1. Growth vectors indicate the response of species 1 (left and right, solid arrows) and species 2 (up and down, dashed arrows) from points a, b, and c. The combined dynamics of the system are given by the sum of the two species' vectors, which will have a diagonal slope. Open circles on the intercepts indicate unstable equilibrium points; closed circles are stable equilibria.

Case II in Figure 4.5 illustrates an arrangement of the isoclines exactly opposite from the example in Case I. Here $K_1 < K_2/\beta$ and $K_2 > K_1/\alpha$, so the effect of species 1 on species 2 is weaker than the effect of species 2 on itself. Additionally, the effect of species 2 on species 1 is stronger than the self-inhibition of species 2. This means that species 2 will eliminate species 1, and the y-intercept at K_2 is a stable equilibrium point.

In Case III, Figure 4.6, the isoclines cross, producing an equilibrium point where both species coexist. In this case, $K_1 < K_2/\beta$ and $K_2 < K_1/\alpha$. By the same reasoning applied to the first two examples, this means that each species inhibits itself more than

Case II

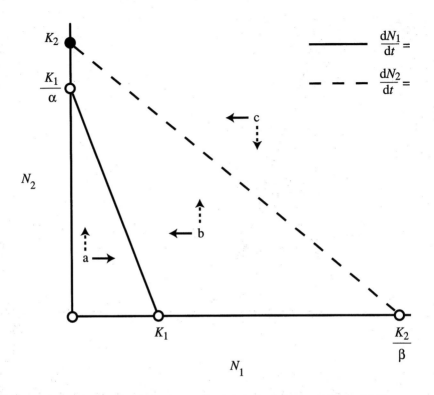

Figure 4.5 **Case II.** This arrangement is exactly opposite from the one in Figure 4.4. Now the species-2 isocline is entirely above and to the right of the species-1 isocline. Species 2 can increase in zone b while species 1 decreases; as a result, species 2 eliminates species 1, growing to its equilibrium density at $N_2 = K_2$.

it inhibits the other. The isoclines in Case III divide the state space into four regions, and population dynamics in all four regions move N_1 and N_2 toward the point where the isoclines cross. This intersection is a globally stable equilibrium. The model thus predicts that when each species limits itself more than it limits the competitor, the interaction will result in coexistence. To find the values of N_1 and N_2 at this joint equilibrium, we solve for the equilibrial densities of each species. Solving equation (4.7) for N_1, the equilibrial density of species 1, yields

$$\hat{N}_1 = K_1 - \alpha N_2 \tag{4.14}$$

The corresponding equilibrial density of species 2, obtained by solving equation (4.11) for N_2, is

$$\hat{N}_2 = K_2 - \beta N_1 \tag{4.15}$$

Case III

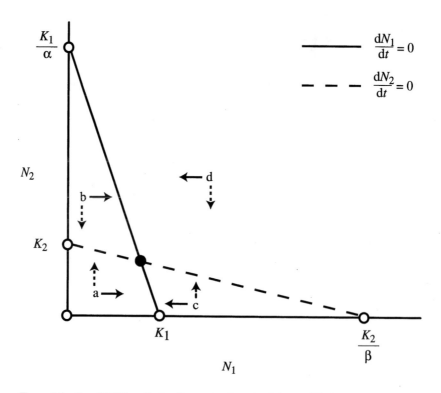

Figure 4.6   **Case III**. When the isoclines cross there is a joint equilibrium point where $dN_1/dt = 0 = dN_2/dt$ and both species coexist. Intercept values show that each species is inhibiting itself more strongly than it inhibits the interspecific competitor because $\alpha\beta < 1$ and therefore $K_1 < K_2/\beta$ and $K_2 < K_1/\alpha$. In zone a combined interference is too low to limit growth of either species, and both will increase in density. In zone b species-2 density falls, but species 1 can still increase. In zone c species-1 density falls but species 2 can still increase. In zone d growth rates of both species are negative. Thus, the interior equilibrium is globally stable and this arrangement results in coexistence.

Now substituting equation (4.15) for N_2 in equation (4.14) and rearranging,

$$\hat{N}_1 = K_1 - \alpha(K_2 - \beta N_1) \tag{4.16}$$

$$\hat{N}_1 = \frac{K_1 - \alpha K_2}{1 - \alpha\beta} \tag{4.17}$$

Substituting equation (4.14) for N_1 in equation (4.15) and rearranging,

$$\hat{N}_2 = K_2 - \beta(K_1 - \alpha N_2) \tag{4.18}$$

$$\hat{N}_2 = \frac{K_2 - \beta K_1}{1 - \alpha\beta} \tag{4.19}$$

Case IV

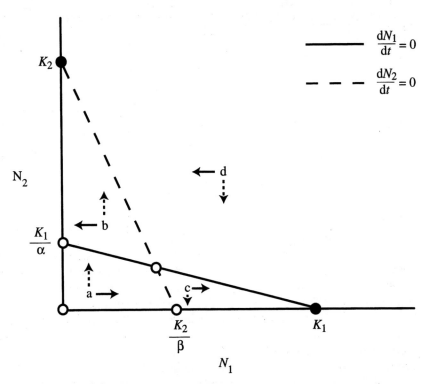

Figure 4.7 **Case IV.** Again the isoclines cross, but in this arrangement the intercept values show that interspecific inhibition is stronger than self inhibition because $K_1 > K_2/\beta$ and $K_2 > K_1/\alpha$. The dynamics in zones a and d are the same as those in Figure 4.6, but in zone b species 2 increases to K_2, while in zone c species 1 increases to K_1. This means that the interior equilibrium is unstable. This interaction results in competitive exclusion, growing either to K_1 or to K_2. The identity of the winner is determined by initial densities and species growth rates.

These $\hat{N}_1$ and $\hat{N}_2$ values are the abundances of species 1 and species 2 at their joint equilibrium. The stability of this joint equilibrium varies with the competition coefficients and carrying capacities. Competition coefficients α and β are usually less than 1. As a result, the denominators of equations (4.17) and (4.19) are usually positive, and the numerators must also be positive to yield positive equilibrial densities of $\hat{N}_1$ and $\hat{N}_2$. This requires that

$$K_2 < \frac{K_1}{\alpha} \quad \text{and} \quad K_1 < \frac{K_1}{\beta} \tag{4.20}$$

conditions that are true when $\alpha\beta < 1$.

The final arrangement, Case IV in Figure 4.7, also has a two-species equilibrium where the isoclines cross. In this case however, $K_1 < K_2/\beta$ and $K_2 < K_1/\alpha$. This means that both interspecific inhibitions are stronger than the intraspecific inhibitions.

Table 4.1

SUMMARY OF THE POSSIBLE OUTCOMES OF LOTKA-VOLTERRA COMPETITION.

Case	Winner	Equilibrial Densities	Stability Criteria
I	1	$N_1 = K_1, N_2 = 0$	$K_1 > \dfrac{K_2}{\beta}$ and $K_2 < \dfrac{K_1}{\alpha}$
II	2	$N_1 = 0, N_2 = K_2$	$K_1 < \dfrac{K_2}{\beta}$ and $K_2 > \dfrac{K_1}{\alpha}$
III	1 and 2	$N_1 = \dfrac{K_1 - \alpha K_2}{1 - \alpha\beta}$ and $N_2 = \dfrac{K_2 - \beta K_1}{1 - \alpha\beta}$	$K_1 < \dfrac{K_2}{\beta}$ and $K_2 < \dfrac{K_1}{\alpha}$
IV	1 or 2	$N_1 = K_1$ or $N_2 = K_2$	$K_1 > \dfrac{K_2}{\beta}$ and $K_2 > \dfrac{K_1}{\alpha}$

The joint equilibrium where the two isoclines cross is unstable, and the outcome of this case is always competitive exclusion. Either species may win, depending on their initial abundances and growth rates. The species with the higher initial abundance and growth rate is more likely to win, but the outcome depends on the actual values of the competition coefficients, carrying capacities, and initial densities. Note that a simulation of Lotka-Volterra competition beginning exactly on the unstable equilibrium densities would continue unchanged, giving the appearance of a stable equilibrium. A more complex formulation including chance effects and environmental variation would not show this unrealistic result (Table 4.1).

All four of these isocline graphs are easy to reproduce using *Populus* simulations. To duplicate them, enter appropriate values for the competition coefficients and carrying capacities (see Box 4.1) and select the N_2 vs N_1 phase-plane output. *Populus* places both the isoclines and the trajectory of changing population densities on the same graph. Figure 4.8 illustrates this output for the default case, which comes to a stable equilibrium at the point where the isoclines cross. From the options menu in an output window, you can also run a stability analysis that initiates N_2 vs N_1 trajectories from many different starting densities.

4.6 CARRYING CAPACITIES AND THE INTENSITY OF COMPETITION AFFECT THE PROBABILITY OF COEXISTENCE

Only one of the isocline arrangements in Figures 4.4–4.7 results in the stable coexistence of competitors. This was Case III in Figure 4.6, where each species' self-inhibition was stronger than the interspecific inhibition it imposed on the competitor. The stability criteria for this case, given by equation (4.20), or by $\alpha\beta < 1$, permit a wide range of numerical values for the individual competition coefficients. Consider an example where the two species are very different so that there is little interspecific inhibition. If α is small, K_1/α will be large and a wide range of K_2 values will satisfy the lefthand inequality of equation (4.20). Similarly, if β is small, K_1/β will be large and a wide range of K_1 values will satisfy the righthand inequality of equation (4.20). We can put this more precisely by combining the two coexistence criteria of equation (4.20):

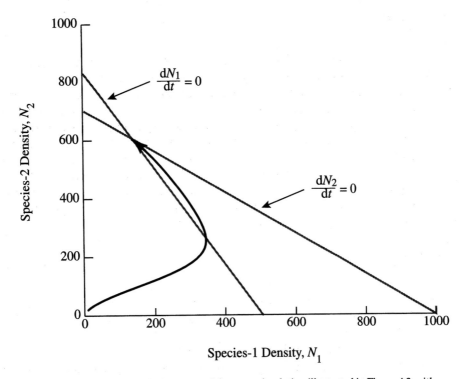

Figure 4.8   *Populus* phase-plane output of the same simulation illustrated in Figure 4.2, with both species' isoclines and the N_2 vs N_1 plot of competitors' population sizes as the interaction proceeds. In this graph, the scales are larger (in comparison with Figure 4.2b) so that all four isocline intercepts fit in the state space.

$$\alpha < \frac{K_1}{K_2} < \frac{1}{\beta} \qquad (4.21)$$

If $\alpha = \beta = 0.1$, then values of the ratio K_1/K_2 between 0.1 and 10.0 will permit coexistence.

Now consider a contrasting example where the two species are very similar. In this case, the competition coefficients α and β will be large (~ 1.0), and the range of K_1 and K_2 values that permit coexistence is much more restricted. Going directly to the combined criteria of equation (5.21), suppose that $\alpha = \beta = 0.9$. Then values of the ratio K_1/K_2 between 0.9 and 1.1 will permit coexistence. Note that this range is much smaller than that of the preceding example.

The conclusion implied by this analysis is that as competition becomes more severe because two competing species have increasingly similar requirements, the probability of competitive exclusion rises. Similarity reaches its logical extreme in the diatom experiment of Figure 4.1, where both algal species have an absolute requirement for silicate, the single limiting resource. Laboratory and field experiments by Gause (1934), Crombie (1945, 1946), Park (1948, 1954), Connell (1961a,b) and others gradually solidified the assertion that two species cannot coexist on a single limiting resource as the competitive exclusion principle. In fact, this idea originated in the theoretical work of Volterra (1926), Lotka (1932), and Gause (1934).

4.7 POSTSCRIPT

In closing a chapter on Lotka-Volterra competition, we should point out that the model is overly simple in at least two ways, and scientists addressing these weaknesses have made important subsequent advances. First, the interaction between organisms is often complex, and to define the *per capita* effect of species 2 on species 1 as a single constant (α) is obviously simplistic. In fact, interspecific feedback may be a complex function of N_2, and environmental influences are left out of the model entirely. A second oversimplification results from quantifying resource availability as a carrying capacity, in units of sustainable consumers. This obscures both the mechanistic details of consumption and the turnover of resources and consumers that mediate competition. A more recent modeling approach that addresses these complications appears in the *Populus* simulation of Resource Competition.

REFERENCES

Brown, J. H., D. W. Davidson, J. C. Munger and R. S. Inouye, "Experimental Community Ecology: The Desert Granivore System." In J. Diamond and T. J. Case (eds), *Community Ecology*. New York: Harper & Row, 1986, 41–62.

Case, T. J., *An Illustrated Guide to Theoretical Ecology*. New York: Oxford University Press, 2000.

Connell, J, "Effects of Competition, Predation by *Thais lapillus* and Other Factors on Natural Populations of the Barnacle *Balanus balanoides*," *Ecol. Mon.*, 1961, 31:61–104.

Connell, J., "The Influence of Inter-Specific Competition and Other Factors on the Distribution of the Barnacle *Chthamalus stellatus*," *Ecology*, 1961, 42:710–723.

Crombie, A. C., "The Competition between Different Species of Graminivorous Insects," *Proc. Roy. Soc. Lond.*, 1945, 132:362–395.

Crombie, A. C., "Further Experiments on Insect Competition," *Proc. Roy. Soc. Lond.*, 1946, 133:76–109.

Gause, G. F., *The Struggle for Existence*. Baltimore: Williams and Wilkins, 1934.

Gotelli, N. J., *A Primer of Ecology*. Sunderland, MA: Sinauer Associates, 1995.

Hutchinson, G. E., *An Introduction to Population Ecology*. Yale University Press, 1978.

Keddy, P. A., *Competition*. New York: Chapman & Hall, 1989.

Lotka, A. J., *Elements of Physical Biology*. Baltimore, Williams and Wilkins, 1925. Reprinted as *Elements of Mathematical Biology*. New York: Dover Publications, Inc., 1956.

Lotka, A. J., "The Growth of Mixed Populations: Two Species Competing for a Common Food Supply." *J. Wash. Acad. Sci.*, 1932, 22:461–469.

Park, T., "Experimental Studies of Interspecies Competition. I. Competition between Populations of the Flour Beetles, *Tribolium confusum* Duvall and *Tribolium castaneum*." *Herbst. Ecol. Mon.*, 1948, 18:267–307.

Park, T., "Experimental Studies of Inter-Specific Competition, II: Temperature, Humidity and Competition in Two Species of *Tribolium*." *Physiol. Zool*, 1954, 27:177–238.

Park, T., "Beetles, Competition, and Populations." *Science*, 1962, 138:1369–1375.

Ricklefs, R. E. and G. L. Miller, *Ecology*, 4th ed. New York: W. H. Freeman and Co., 1999.

Roughgarden, J., *Theory of Population Genetics and Evolutionary Ecology: An Introduction*, New York: McMillan-Collier, 1979.

Tansley, A. G., "On Competition between *Ganium saxatile* L. (*G. Hercynicum* Weig.) and *Galium sylvestre* Poll (*G. asperum* Schreb.) on Different Types of Soil." *J. Ecol.*, 1917, 5:173–179.

Tilman, D., M. Mattson and S. Langer, "Competition and Nutrient Kinetics along a Temperature Gradient: An Experimental Test of a Mechanistic Approach to Niche Theory." *Limnol. Oceanogr.* 1981, 26(6):1020–1033.

Vandermeer, J. H., "The Community Matrix and the Number of Species in a Community." *American Naturalist*, 1970, 104:73–83.

Volterra, V., "Variazioni e Fluttuazioni del Numero D'individui in Specie Animali Conviventi." *Mem. R. Acad. Naz. Dei Licei.* 1926, 2:31–113. (English translation by M.E. Wells) In *Animal Ecology.* McGraw-Hill, 1931.

PROBLEMS AND EXERCISES

1. Select Lotka-Volterra Competition from the Models menu of *Populus* to display the input window, which is also shown in Box 4.1. Change the initial abundance of species 2 to $N_2 = 0$ so that only species 1 is present, and reduce the growth rate of species 1 to $r_1 = 0.1$. Then, press VIEW to display the default N vs t output graph. Is this a familiar graph? Explain the biology of this result, including definitions of r_1, K_1, r_2, and K_2. Can you show mathematically why the Lotka-Volterra competition model gives this result when the initial abundance of species 2 is $N_2 = 0$?

2. Notice that the *Populus* simulation of Lotka-Volterra Competition allows you to view two different output plots. An N vs t trajectory shows changing densities of both competing species as a function of time. An N_2 vs N_1 phase-plane diagram allows isocline plots like Figures 4.4–4.7. The program also traces a trajectory of population density values on the phase diagram (Figure 4.8), beginning with the initial densities specified on the input screen, and proceeding in the direction indicated by an arrowhead. Reset *Populus* to the default parameter values and open windows to display both output graphs. You can run a sensitivity analysis on the N_2 vs N_1 graph using the cursor to specify any set of starting densities for a new trajectory. The program will also initiate trajectories from points around the perimeter of the plot space. Compare the N vs t trajectories and N_2 vs N_1 phase-plane diagrams for several simulations with different parameter values until you are comfortable with the relationship between the two output plots.

3. Figures 4.4–4.7 introduced four cases that predicted different competitive outcomes based on the relative positions of the species-1 and species-2 isoclines. Which of the four cases is represented by the *Populus* default parameter values? What outcome does this case predict, and what are the criteria (relative values of the carrying capacities and competition coefficients) that cause this result? Explain the outcome and biological interpretation of all four cases in your own words.

4. Now experiment with other values of the input parameters, and observe their effects. Try varying the competition coefficients (α and β) and carrying capacities (K_1 and K_2) to produce examples of all four cases in Figures 4.4–4.7. As you

do so, fill in values for the four cases in the table below. Start by putting the default values in the appropriate column. Run each simulation, observing both the N_2 vs N_1 plots and the N vs t plots. Make sure that you can predict what the N vs t plot will look like from knowledge of the N_2 vs N_1 plot, and vice versa.

	Case I	Case II	Case III	Case IV
$N_1(0)$				
$N_2(0)$				
r_1				
r_2				
K_1				
K_2				
α				
β				

5. In which of the four cases listed above is the outcome of competition most likely to vary with the values of r_1 and r_2? Justify your answer by choosing the appropriate case-set of parameter values from your table in Problem 4, and specifying two pairs of r values that produce different competitive outcomes.

$r_1 =$ $r_2 =$ winner =

$r_1 =$ $r_2 =$ winner =

6. In which of the four cases listed above is the outcome of competition likely to vary with the initial population densities of the two competing species? Justify your answer by choosing the appropriate case-set of parameter values from your table in Problem 4, and specifying two pairs of initial values for N_1 and N_2 that produce different competitive outcomes.

$N_1(0) =$ $N_2(0) =$ winner =

$N_1(0) =$ $N_2(0) =$ winner =

7. What is the range of values for the competition coefficients, α and β, that permit the stable coexistence of Lotka-Volterra competitors? Explain in your own words why values outside this range result in competitive displacement.

8. (a) Consider two cases of competitive interaction between species with very similar carrying capacities in an experimental habitat (e.g., $K_1 = 500$, $K_2 = 505$). For the first case, assume that resource requirements for the two species overlap very little, so that competition is relatively weak (e.g., $\alpha = \beta = 0.1$). For the second case, assume that overlap is large and competition is strong (e.g., $\alpha = \beta = 0.9$). How will the isocline plots differ? Sketch them and then use *Populus* simulations to check your answer.

(b) Now consider examples with moderate overlap between species (e.g., a = b = 0.40), varying the difference between carrying capacities (e.g., try K1 = 500, K2 = 700, and then K1 = 200, K1 = 700). How will the isocline plots differ? Sketch them and use Populus simulations to check your answer.

(c) What would happen if $K_1 = K_2$, and $\alpha = \beta = 1.0$? How would this affect the isoclines? What dynamics would you expect from such a system?

PROBLEMS WITH EMPIRICAL DATA

9.   Data reproduced in the following table come from a laboratory competition experiment by Thomas Park (1948) cited in the reference section (specifically, these data
are from Park's Tables 3, 6, and 16). Park ran replicate colonies of the flour beetles
Tribolium castaneum and *Tribolium confusum* in bottles with 40 grams of flour.
Beetles, beetle larvae and eggs were sieved from the flour, counted, and placed on
new flour at 30-day intervals. All cultures were initiated with 1 beetle per gram of
flour and equal numbers of males and females. Control cultures received one beetle species or the other, while experimental cultures received equal numbers of
both species. Do Park's data show evidence of competition between *Tribolium castaneum* and *Tribolium confusum*? Graph the population densities of each species
in both control and experimental cultures over the first year or more of the experiment, making both N vs t trajectories and phase-plane plots. Which species has the
higher intrinsic growth rate under these culture conditions? Which species has the
higher carrying capacity? From Park's data, can you judge the relative intensities of
interspecific and intraspecific interference for each species? Which of the Lotka-
Volterra cases in Figures 4.4–4.7 best represents the beetle interaction?

Culture Age (days)	Control T. confusum	Control T. castaneum	Experimental T. confusum	Experimental T. castaneum
30	27.2	23.4	10.2	11.0
60	27.7	27.7	6.1	18.0
90	24.8	30.0	5.0	20.2
120	23.4	26.1	4.6	17.7
150	22.3	21.4	4.2	16.0
180	20.7	16.0	3.9	11.4
210	19.3	7.6	4.4	5.8
240	18.8	10.8	5.7	6.7
270	16.6	6.9	7.8	5.5
300	15.8	7.6	10.4	3.3
330	15.9	9.4	12.4	2.2
360	14.8	10.6	10.6	1.4
390	14.8	10.9	11.6	1.7
420	14.9	10.0	9.0	2.2
450	15.7	10.5	11.1	3.0
480	15.4	8.8	12.2	2.2
510	15.6	8.7	11.2	1.6
540	15.7	9.1	11.5	1.3
570	15.5	8.6	11.1	1.2
600	14.1	8.9	11.0	1.6
630	14.2	8.5	11.5	1.0
660	15.6	7.1	12.2	1.8
690	15.4	8.2	8.5	4.0
720	16.4	7.9	10.1	4.0
750	17.5	9.2	10.8	2.9
780	16.4	9.5	12.6	2.6
810	15.4	8.6	13.9	2.3
840	14.8	7.2	12.9	1.1
870	15.4	8.0	9.9	0.9
900	15.9	9.7	14.9	1.2
930	17.7	11.2	16.3	0.3

| 960 | 17.3 | 10.9 | 15.1 | 0.2 |
| 990 | 17.4 | 19.2 | 15.8 | 0.1 |

10.   Review the diatom experiment described at the beginning of this chapter and the data in Figure 4.1. What inferences can you draw about Lotka-Volterra parameter values from these empirical results? Estimate a value for each of the parameters listed in problem 4, explaining both the basis and certainty of your estimate. Then use these values in a *Populus* simulation of Lotka-Volterra competition between *Synedra* and *Asterionella*. Does your simulation produce the same outcome as the lab experiment? Finally, explain the sensitivity of the simulation result. Which parameter values are critical to winning or losing?

FOR ADVANCED STUDENTS

11.   Consider a Lotka-Volterra simulation that results in competitive exclusion with the same winner from any set of beginning densities. Is it possible to convert this example into one that yields stable coexistence by changing only the carrying capacities, K_1 and K_2? If you need a hint, study the graphical treatment of this issue by Vandermeer (1970) or its recapitulation by Keddy (1989). Can you explain the biology of this result using the mathematical criteria for coexistence and exclusion in Table 4.1?

CHAPTER 5
Continuous Predator-Prey Models

Predators and prey have opposing effects on their respective population dynamics. Predators prosper when prey are common, increasing their negative effect on prey density. Conversely, if predators suffer lower survival and reproduction when prey are rare, declining predation should promote prey recovery. Because these adjustments result from changes in population size by both predators and prey, they are not instantaneous. There are likely to be lags associated with population growth and decline that will cause dynamic cycles of higher and lower density in this coupled system. Figure 5.1 shows a famous empirical example. These remarkable data on lynx and hare dynamics

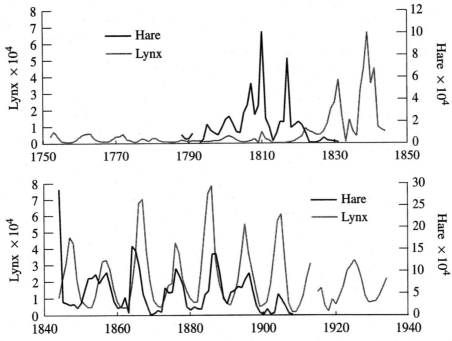

Figure 5.1   Canadian fur records of the Hudson's Bay Company showing cycles in the abundance of snowshoe hare (*Lepus americanus*) and lynx (*Lynx canadensis*) that depend on them for food. The data reflect influences in addition to the underlying biological cycle; low returns from 1778 to 1790 correspond with a smallpox pandemic among the trapping tribes, and concurrent westward expansion of Canadian commercial activity is obvious in the total volume of trade. These data come from Poland (1892), supplemented by Seton (1911), Hewitt (1921), MacLulich (1937) and Elton and Nicholson (1944).

come from Canadian trapping records of the Hudson's Bay Company, which recorded the number of pelts shipped to England annually for nearly two centuries. They suggest that there have been sustained oscillations in the density of lynx and hares, with roughly constant period (time between peaks) and amplitude (difference between peaks and troughs), throughout recorded history.

Lynx specialize almost exclusively on snowshoe hares as prey, so it is not surprising that their populations decline when hares are uncommon. What is less clear is whether lynx could ever limit the prodigious reproductive capacity of hares, causing oscillations without an additional source of density-dependent prey mortality, like disease, or some corresponding cycle in the vegetation consumed by hares. In this chapter, we will examine two continuous, differential-equation models of predator-prey dynamics that formalize the intuitive expectation of cycles and allow a detailed analysis.

5.1 THE LOTKA-VOLTERRA PREDATOR-PREY MODEL

A predator-prey model requires equations for the dynamics of both predator and prey, and each of these equations needs at least two terms. The prey equation will have a positive population-growth term, and depending on our assumptions, it may have a negative term for intraspecific, density-dependent feedback. Because predators consume some prey, the prey equation must also have a negative term modeling this interspecific effect. In the same manner, a predator equation needs an intraspecific term for the dynamics of predators without prey. For simplicity, we will model a tightly coupled system with a highly specialized predator like the lynx. Without prey, these predators will be starving, so the intraspecific term in our predator equation will be negative, dominated by a death rate. There will also be a positive interspecific term indicating the effect of prey consumption, which provides predators with the means for survival and reproduction.

The form and complexity of predator-prey equations depend on our assumptions. We will begin with a highly simplified model developed independently by Lotka (1925) and Volterra (1926), and then examine the effect of more realistic complications. Specifically, let's assume the following:

a. Except for the presence of predators, prey live in an ideal environment. This allows us to use a density-independent model of prey growth, without a term for intraspecific, density-dependent feedback.

b. Predators depend on a single species of prey, and the number of surviving offspring produced by predators is directly proportional to the number of prey they consume. Predator population density does not affect an individual predator's chances of survival and reproduction directly; the only negative feedback on predators is indirect, mediated by effects on prey population size.

c. Both populations reproduce continuously, producing identical individuals that are ageless and sexless. This eliminates the potential complications of age structure, mating, and evolution, and allows us to phrase the model in a pair of continuous differential equations.

d. Finally, we assume that predation occurs in proportion to the number of encounters between predators and prey, without time lags. Here we avoid considering predator satiation, predator and prey life histories, movement, hiding, and hunting behavior.

We will begin our description of the Lotka-Volterra predator-prey model by specifying the dynamics of the predator and prey populations living separately. In other words, we will begin with the intraspecific components of the model. Let N represent the population density of prey, and P the density of predators, where density is the population size per unit area. What is the growth rate of each party in the absence of the other? We assume prey to have an ideal environment, so in the absence of predators their population will grow exponentially, as described in Chapter 1.

$$\frac{dN}{dt} = r_1 N \tag{5.1}$$

Here r_1 is the intrinsic, *per capita* growth rate of the prey. Recall that r has two components, $r = b - d$, representing the *per capita* birth and death rates. Since we assume that predator births occur in proportion to the number of prey consumed, predators without prey will have no progeny, $b_2 = 0$. The *per capita* growth rate, r_2, of these starving predators will have only one component, so $r_2 = -d_2$. Their overall dynamics will be a negative function of population size, resulting in an exponential predator decline.

$$\frac{dP}{dt} = -d_2 P \tag{5.2}$$

Now we need to combine the predators and prey, adding interspecific terms that have positive and negative effects on the densities of predator and prey, respectively. We assume that the number of prey eaten is proportional to the number of encounters between predators and prey. Let C be a proportionality constant specifying what fraction of encounters result in prey consumption. We will assume further that these encounters occur randomly, increasing with the product of predator and prey densities, NP. Then prey consumption will occur in direct proportion to the product of their densities, CNP. Combining intraspecific and interspecific terms, the prey equation is now

$$\frac{dN}{dt} = r_1 N - CNP \tag{5.3}$$

The product CN in equation (5.3) is the predators' **functional response** defining their *per capita* rate of prey consumption as a function of prey density. In this example, the functional response is linear; the number of prey consumed per predator rises in direct proportion (C) to prey abundance (N). This is a simple, but unrealistic formulation of the functional response. It implies that predators with large numbers of prey available could actually eat large numbers, when they are more likely to become satiated. The functional response encapsulates attributes of both predator and prey biology. It is affected by prey escape ability, and by the time predators require to subdue and consume prey before beginning to hunt again. Later in the chapter, we will analyze some of these complications with more realistic formulations.

Prey consumption has a positive effect on predator dynamics, so the same CNP term from equation (5.3) will appear in the predator equation with a positive sign. In this case, however, it signifies the number of new predators produced as a result of eating CNP prey, so we need one more parameter, g, to indicate the rate at which prey are converted into predators. Combining this interaction term with the predator-only model of equation (5.2),

$$\frac{dP}{dt} = -d_2 P + gCNP \tag{5.4}$$

The product gCN in equation 5.4 is the predators' **numerical response**, defining the rate at which new predators are born as a function of prey density. It encapsulates the same biological phenomena as the functional response but depends also on the predators' efficiency at converting prey into new predators.

Together, these prey (5.3) and predator (5.4) equations comprise the Lotka-Volterra predator-prey model. The prey equation (5.3) has a positive intraspecific term for exponential growth in the absence of predators, and a negative interspecific term subtracting prey killed by the predator. The predator equation (5.4) has a negative intraspecific term for predators starving without prey, and a positive interspecific term for the effect of prey consumption on predator dynamics.

DYNAMICS OF LOTKA-VOLTERRA PREDATORS AND PREY

Like the Lotka-Volterra-competition equations of Chapter 4, the dynamics of these predator-prey equations are intertwined. There is no definite-integral solution projecting either the predator or prey populations independently into the future. To study their dynamics, we will use numerical integration to sum successive changes in predator and prey density with a computer and isocline analyses of model equilibria.

The *Populus* simulation of Lotka-Volterra Predators and Prey appears as an option under "Continuous Predator Prey Models" on the "models" menu. Figure 5.2 illustrates a *Populus* simulation based on the default parameter values of Box 5.1. The upper graph is a *Populus* time trajectory, with N and P plotted against t. This numerical integration of equations (5.3) and (5.4) shows sustained oscillations in predator and prey density, similar to the lynx-hare data. The lower graph is a phase-plane representation of the same simulation, with P on the y-axis and N on the x-axis. Notice that the sustained oscillation retraces the same predator and prey densities, forming a closed loop in the phase-plane plot.

ISOCLINE ANALYSES

We can also analyze the joint equilibrium of the Lotka-Volterra Predator-Prey equations, using the isocline techniques developed in Chapter 4. First, we derive a prey isocline, setting the prey growth equation (5.3) equal to zero and rearranging:

$$\frac{dN}{dt} = 0 = N(r_1 - CP) \tag{5.5}$$

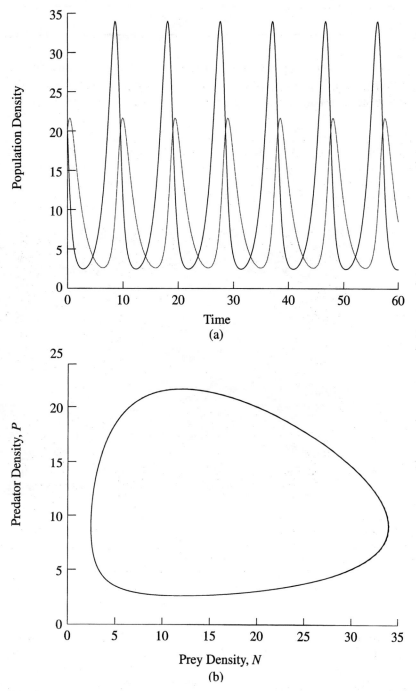

Figure 5.2   *Populus* simulation of a Lotka-Volterra predator-prey interaction based on the default parameter values of Box 5.1 ($N_0 = P_0 = 20, r_1 = 0.9, C = 0.1, d_2 = 0.6, g = 0.5$). The upper graph plots a time trajectory of prey and predator population sizes, N and P, with cycles resembling the lynx-hare data. The lower graph plots the same simulation on an N–P phase plane. Here the cycles form a closed loop that is retraced with each new cycle of prey and predator abundances.

BOX 5.1

The *Populus* input window for a Lotka-Volterra predator-prey simulation, with parameter values and option settings that produce the oscillating densities of Figure 5.2a. Selecting the *P* vs *N* plot-type option gives the phase-plane output graph shown in Figure 5.2b.

The Lotka-Volterra model is an option under Continuous Predator-Prey Models. It runs with or without density-dependent feedback on prey-population growth and provides both time trajectories and phase-plane output graphs.

The Lotka-Volterra button in the model-type box brings up a tabbed parameter sheet. The carrying capacity window in the prey box is only accessible when the density-dependent prey option is invoked.

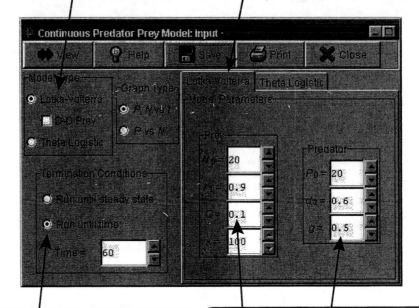

The default termination option for this module is a fixed runtime. The model only comes to a stable equilibrium with density-dependent prey.

C sets the proportion of predator-prey encounters resulting in prey mortality; *g* is the conversion rate of captured prey into new predators.

Because the righthand side of this equation is the product of two terms, there are two solutions that fulfill the equality. The solution where $N = 0$ is trivial; a prey population cannot increase if there are no prey. The interesting case is the one where the parenthetic term $(r_1 - CP)$ equals zero. This gives the number of predators required to hold prey population density at some constant, positive value.

$$0 = r_1 - CP \qquad (5.6)$$

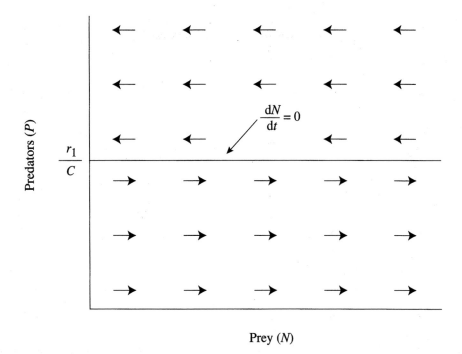

Figure 5.3   Zero net growth isocline for Lotka-Volterra prey. The flat line marks the constant predator density (r_1/C) that holds $dN/dt = 0$. In the zone below the isocline, there are too few predators to prevent prey density from increasing. Above the isocline, there are enough predators to cause a prey-population decline.

$$P = \frac{r_1}{C} \tag{5.7}$$

Equation (5.7) tells us that a constant density of predators, r_1/C, will hold $dN/dt = 0$. This predator density is directly proportional to the prey growth rate, r_1, and inversely proportional to the probability of prey mortality per predator-prey encounter, C. The prey isocline shown in Figure 5.3 is flat; if predator density is above the isocline, then the prey population will shrink. If predator density is below the isocline, then the prey population will grow. This peculiar result rests on two of our initial assumptions. The first is the assumption of density independence. Prey that suffer density-dependent reductions of *per capita* survival and reproduction would be increasingly limited by intraspecific feedback as they approach their carrying capacity and would require fewer predators to hold $dN/dt = 0$. In this case, the prey isocline would slope downward, with an x-intercept at $N = K$. The second assumption responsible for the flat isocline of Figure 5.3 is the linear functional response. If predators become satiated at high prey density, more predators would be required to hold $dN/dt = 0$ as prey density rises, and the isocline would slope upward. Both assumptions are unrealistic features of the Lotka-Volterra model that we will explore later in the chapter.

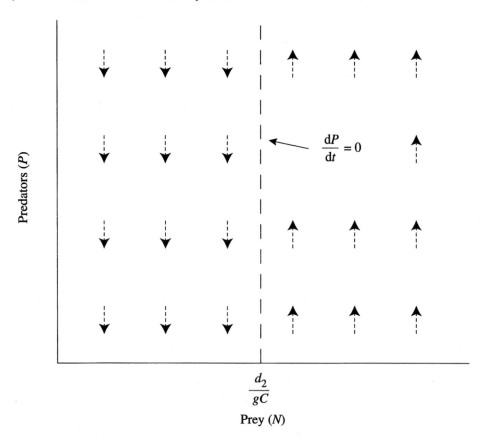

Figure 5.4 Zero net growth isocline for Lotka-Volterra predators. The vertical line marks the prey density (d_2/gC), that holds $dP/dt = 0$. In the zone to the left of the isocline, there are too few prey to maintain the predators at constant density, and the predator population declines. To the right of the isocline, there are sufficient prey available to allow a predator-population increase.

Now we use the same manipulations to derive an isocline for predator equation (5.4).

$$\frac{dP}{dt} = 0 = -d_2P + gCNP \tag{5.8}$$

$$0 = P(gCN - d_2) \tag{5.9}$$

Again, we will consider the case in which the parenthetic term equals zero.

$$0 = gCN - d_2 \tag{5.10}$$

$$N = \frac{d_2}{gC} \tag{5.11}$$

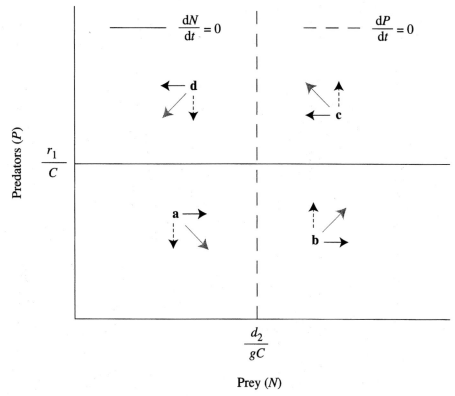

Figure 5.5   Lotka-Volterra isoclines for both predators and prey superimposed on the N–P plane. Dynamics in the four resulting quadrants are given by the predator (vertical) and prey (horizontal) growth vectors. Gray vectors represent the sum of predator and prey components, illustrating the model's counterclockwise cycling.

Thus a constant number of prey, d_2/gC, defines the predator isocline. Graphing this isocline with prey-population density on the x-axis and predator-population density on the y-axis gives the vertical line of (Figure 5.4). If prey density is lower than d_2/gC, predators do not find enough food to replace themselves, and predator density falls. If prey density is above d_2/gC, the predator birth rate exceeds the death rate, and predator-population density increases.

Figure 5.5 superimposes the predator and prey isoclines so that we can examine the dynamics of both species simultaneously. The isoclines cross at an equilibrium density for both species. To investigate whether this joint equilibrium might be stable or unstable, we inspect the four quadrants surrounding the intersection. The predator and prey growth vectors direct densities down and to the right in quadrant **a**, up and to the right in **b**, and so on; the summation vectors point neither toward nor away from the joint equilibrium. Instead, they are roughly perpendicular to the direction of the crossed isoclines, defining a loop like the one shown in Figure 5.2b. Figure 5.6 plots three Lotka-Volterra predator-prey time trajectories that begin with different initial densities, N_0 and P_0, but share all other parameter values. Both predators and

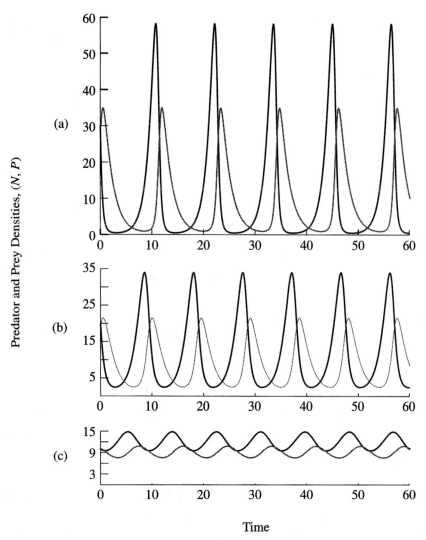

Figure 5.6   *Populus* simulations of the Lotka-Volterra predator-prey model. Black lines give the prey trajectory; gray lines represent the predator. In all three graphs, parameter values for the prey were $r_1 = 0.9$, and $C = 0.1$. For the predators, $d_2 = 0.6$, $g = 0.5$. Initial population densities for graph (a) were $N_0 = P_0 = 30$; for (b) $N_0 = P_0 = 20$ (this case duplicates Figure 5.2); and for (c) $N_0 = P_0 = 30$. The same three simulations are plotted with isoclines on the N–P phase plane in Figure 5.7.

prey oscillate in numbers, with periods and amplitudes set by the initial densities. Figure 5.7 gives the corresponding isocline plot on an N–P phase plane. The three simulations with different initial densities form a series of concentric figures corresponding to the time trajectories of differing period and amplitude.

We call an oscillating model stable if fluctuations damp out and population sizes converge to constant values. In contrast, fluctuations are unstable if they get larger and larger until either predator or prey become extinct. The dynamics of a Lotka-Volterra predator-prey system are **neutrally stable**. If such a system begins exactly at the joint

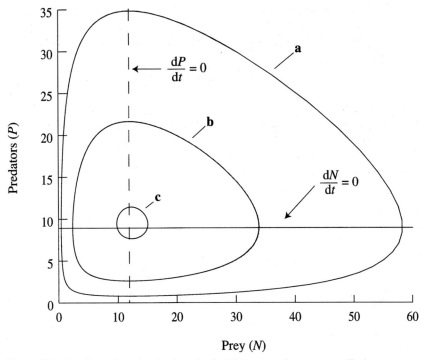

Figure 5.7   *Populus* trajectories for three Lotka-Volterra predator-prey oscillations graphed with isoclines on the *N–P* plane. All parameter values are those given for Figure 5.6. Neutrally stable oscillations retrace the same trajectory on each successive cycle. The period and amplitude of the cycles are determined by initial population sizes, the predator and prey growth rates, and the functional and numerical responses.

equilibrium where the isoclines cross, it remains there. If it begins with some other densities, the populations oscillate with a constant period and amplitude that depends on those initial densities. Neutrally stable oscillations have no tendency to get larger or smaller over time.

How well does the Lotka-Volterra model describe the dynamics of lynx and hares? At first glance, the fit is not too bad; there are oscillations of roughly constant period and amplitude in Figure 5.6 and in the 1850–1910 interval of Figure 5.1. However, constant period and amplitude are not what we should expect from Lotka-Volterra predators and prey in the real world. Since the model is neutrally stable, any environmental perturbation that affects population size should establish cycles of a new period and amplitude. Thus the relative constancy of the lynx-hare cycle is actually troublesome; it does not show the influence of environmental variation that we would expect in a Lotka-Volterra predator-prey system.

MODIFYING THE LOTKA-VOLTERRA MODEL WITH DENSITY-DEPENDENT PREY

When a theoretical model is not consistent with our empirical data, we need to examine its underlying assumptions. One assumption of the basic Lotka-Volterra model that is clearly unrealistic is its density-independent prey. We can examine the effect of density dependence on predator-prey dynamics by adding intraspecific feedback to the prey equation (5.3) in the form of a carrying capacity, K.

$$\frac{dN}{dt} = r_1 N\left(1 - \frac{N}{K}\right) - CNP \tag{5.12}$$

Now a negative feedback term $1 - N/K$ slows prey population growth as density approaches the carrying capacity. In the absence of predators, the interspecific term, $-CNP$, would drop out, leaving a prey population with logistic growth. Setting equation (5.12) equal to zero, we can derive a new prey isocline incorporating this density-dependent feedback.

$$\frac{dN}{dt} = 0 = r_1 N\left(1 - \frac{N}{K}\right) - CNP \tag{5.13}$$

$$0 = N\left[r_1\left(1 - \frac{N}{K}\right) - CP\right] \tag{5.14}$$

The solution with $N = 0$ is uninteresting, so we will examine the bracketed term.

$$0 = r_1\left(1 - \frac{N}{K}\right) - CP \tag{5.15}$$

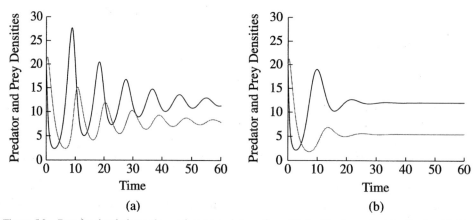

Figure 5.8 *Populus* simulations of a predator-prey interaction with density-dependent feedback added to the prey equation of the basic Lotka-Volterra model. Black lines give the prey trajectory; gray lines represent the predator. For the simulation on the left, parameter values for predators were $P_0 = 20$, $d_2 = 0.6$, $C = 0.1$, $g = 0.5$. For prey, $N_0 = 20$, $r_1 = 0.9$, and $K = 100$. Density dependence changes the neutrally stable Lotka-Volterra cycle to a damped oscillation that converges on stable equilibrium densities. For the simulation on the right, all parameter values are identical except that the prey carrying capacity is reduced from $K = 100$ to $K = 30$. This increases density-dependent feedback and causes the dynamics to converge more quickly to a somewhat lower predator density. The same trajectories are graphed with isoclines on the $N-P$ plane in Figure 5.9.

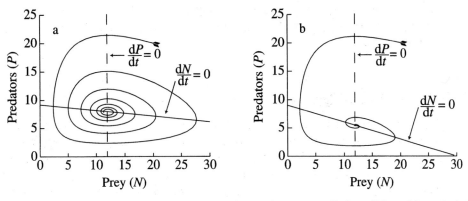

Figure 5.9   *Populus* trajectories for the density-dependent predator-prey oscillations of Figure 5.8, graphed with isoclines on the *N–P* plane. Increased density-dependent feedback in the case on the right makes the prey isocline slope more negative, causing predator-prey oscillations to converge more quickly to equilibrial densities.

$$CP = r_1 \left(1 - \frac{N}{K} \right) \tag{5.16}$$

$$P = \frac{r_1}{C} - \frac{r_1}{CK} N \tag{5.17}$$

The isocline given by equation (5.17) has the same *y*-intercept, r_1/C, as the density-independent version in equation (5.7), but it is no longer flat, tilting downward with a slope of $-r_1/CK$. Prey are increasingly limited by intraspecific, density-dependent feedback as their density rises. Thus the predator density required to hold $dN/dt = 0$ declines in proportion to prey density, *N*. At the *x*-intercept where $P = 0$, prey regulation results entirely from intraspecific feedback and $N = K$. The dynamic effect of density dependence converts the neutrally stable Lotka-Volterra oscillation of Figures 5.6 and 5.7 to a stable, damped oscillation in Figures 5.8 and 5.9. The prey isocline of Figure 5.9a shows the negative slope conferred by the addition of density-dependent feedback. If the isocline slope is more negative, the oscillation converges more rapidly and there are fewer cycles before the interaction reaches equilibrium (Figure 5.9b).

To calculate the predator and prey densities at their joint equilibrium, we begin with the predator isocline, equation (5.11). The number of prey required to hold $dP/dt = 0$ is $\hat{N} = d_2/C$. Substituting this value into equation (5.17) gives the equilibrial predator density, *P*.

$$\hat{P} = \frac{r_1}{C} - \frac{r_1}{CK} \frac{d_2}{C} \tag{5.18}$$

$$\hat{P} = \frac{r_1}{C} \left(1 - \frac{d_2}{CK} \right) \tag{5.19}$$

One motivation for adding density-dependent feedback to our analysis was to see whether it would improve the fit of our basic Lotka-Volterra model to the lynx-hare data of Figure 5.1. Clearly this is not the case. Density-dependent prey growth damps a predator-prey oscillation, causing the cycles to converge toward stable equilibrial densities.

5.2 THE θ–LOGISTIC MODEL

There is another simplifying assumption of the Lotka-Volterra predator-prey model that is both important and unrealistic; this is the assertion that insatiable predators consume prey in direct proportion to their encounter rate. Imagine that you have just finished eating a huge holiday dinner. The assumption that predators consume a constant proportion of the available prey, CN, implies that with ten times more food on the table, you could have eaten ten times more in the same amount of time. Even if you did not get full (and need a nap!), there is a limit to the rate at which food can be consumed. In this section, we analyze a θ-Logistic Predator-Prey Model (θ is the Greek letter theta). This generalized relative of the basic Lotka-Volterra model allows more complex and realistic functional responses.

We will begin with a partial implementation of the θ-Logistic Model focusing on the predators' functional response. Consider this equation for the change in prey population size with time:

$$\frac{dN}{dt} = r_1 N \left(1 - \frac{N}{K} \right) - f(N)P \tag{5.20}$$

It is directly analogous to equation (5.12). The intraspecific component has a negative feedback term, $1 - N/K$; in the absence of predators these prey would grow logistically. The interspecific term has now been generalized so that the functional response (the predators' *per capita* consumption of prey) can be any function, $f(N)$, of prey density. For equation (5.12) this function was a constant proportion of the prey population, $f(N) = CN$, with the unrealistic consequence that individual predators could consume extremely large numbers of high-density prey. On the following pages we will substitute several different formulations for $f(N)$ to explore satiation and other aspects of predator foraging behavior.

We will specify the predator dynamics for our θ-logistic model as follows:

$$\frac{dP}{dt} = gP[f(N) - D] \tag{5.21}$$

Here $f(N)$ indicates the predators' functional response, as in equation (5.20). D can be interpreted as the minimum *per capita* rate of prey consumption required for a predator to replace itself in the next generation, and g sets the conversion efficiency with which prey consumption is translated into new predators. The bracketed term of (5.21) is the net prey consumption (the difference between gross prey capture and the break-even requirement) per predator at prey density N. This means that predator-population growth is proportional to the number of predators and their net prey intake *per capita*.

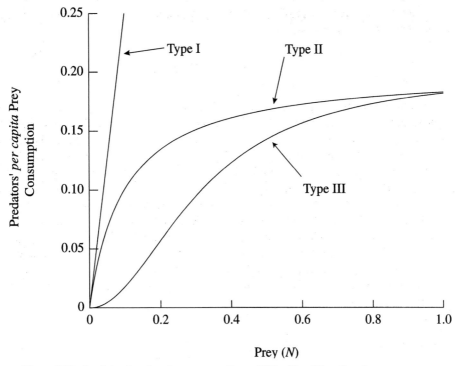

Figure 5.10 Predator functional response patterns. With a Type I functional response, prey consumption rises in direct proportion to prey density. The Type II response is a saturating function with an upper bound, imposed either by predator satiation or because handling consumes a larger fraction of the predator time budget as prey density increases. For Type III predators, prey consumption is a second-order function of prey density. The capture rate *per capita* first accelerates and then saturates at an upper limit as prey density increases.

FUNCTIONAL RESPONSES

Because a variety of expressions can be incorporated for f(N) in both the predator and prey equations, the θ-logistic model lets us compare the dynamic effects of different functional responses. The functional responses of many predator species have been the subject of laboratory experiments. For example, Canadian ecologist C. S. Holling placed predatory mice in an arena with different numbers of insect pupae (1965) and praying mantis predators with housefly prey (1966) to see how the predators' *per capita* consumption varied with prey density. Three patterns (shown in Figure 5.10) result from such experiments (Hassell 1978). The Type I response rises linearly in direct proportion to prey density, so

$$f(N) = CN \qquad (5.22)$$

This is the functional response incorporated in the Lotka-Volterra model.

The Type II response rises at a continually decreasing rate, so that prey capture reaches an asymptotic upper limit. Several mathematical expressions can represent this response; the form used in *Populus* is

$$f(N) = \frac{CN}{1 + hCN} \tag{5.23}$$

As before, C is the proportion of encountered prey that are captured per predator, per unit time. The new parameter, h, is handling time, the average time a predator requires to subdue and consume each prey, prior to the resumption of searching. Time spent handling prey is not available for searching. As the number of available prey (N) becomes large, the $f(N)$ given by equation (5.23) approaches an asymptotic value of $1/h$ because the predator spends most of its time handling prey and little time searching for new prey. This assumes that prey are never captured while another prey item is being handled, and prey are captured at rate CN while the predator is searching. A Type II functional response would also result if predators search only when they are hungry.

The Type III response is sigmoid, rising slowly when prey are rare, accelerating when they become more abundant, and finally reaching a saturated upper limit like the Type II curve. The most common mathematical representation of a Type III functional response is

$$f(N) = \frac{CN^2}{1 + hCN^2} \tag{5.24}$$

Here the same interpretation of h as a handling time is possible, but predators catch prey at the rate CN^2 while searching. Type III responses might occur if a predator learns to handle prey more efficiently as they become more abundant or switches from rare to common prey types as their relative densities fluctuate (Murdoch 1969).

THE θ-LOGISTIC MODEL WITH A TYPE II FUNCTIONAL RESPONSE

The saturation of prey consumption, either because the predators' time is entirely spent handling prey or because their appetite is satiated, implies a positive density-dependent feedback of prey density on prey growth. Predators become less effective in controlling prey as prey density increases. This suggests that simulations with a Type II functional response should be less stable than those resulting from a Type I model like the density-dependent Lotka Volterra model of equations (5.12) and (5.4). We can test this expectation with an isocline analysis. Substituting the Type II functional response [equation (5.23)] for $f(N)$ in the θ-logistic prey equation of (5.20), we have

$$\frac{dN}{dt} = r_1 N \left(1 - \frac{N}{K} \right) - \frac{CNP}{1 + hCN} \tag{5.25}$$

After setting equation (5.25) equal to zero, several algebraic rearrangements yield the following prey isocline:

$$P = -\frac{r_1 h}{K}N^2 + \left(r_1 h - \frac{r_1}{C}\right)N + \frac{r_1}{C} \tag{5.26}$$

Instead of a line this second-order function is curved, with y-intercept (where $N = 0$) of $P = r_1/C$, and x-intercept (where $P = 0$) of $N = K$ (*cf.* Figure 5.12b, which shows a θ-logistic simulation based on the default parameter values of Box 5.2. The isocline is humped, with a positive slope near its y-intercept, because each predator consumes a

---

BOX 5.2

The *Populus* input window for a θ-Logistic predator-prey simulation, with parameter values and option settings that produce the sustained limit cycle in Figure 5.11b. Selecting the P vs N plot-type option gives the phase-plane output graph shown in Figure 5.12b. Figures 5.11a and 5.12a illustrate a corresponding example with a Type I functional response.

The θ-logistic model is an alternative on the same input window as the Lotka-Volterra model. It allows both linear and saturated functional responses, and nonlinear density-dependent feedback on prey population growth.

The θ-Logistic button in the model-type box brings up a tabbed parameter sheet, including functional response parameters that are not used in the basic Lotka-Volterra model. The default outputs are shown in Figures 5.11b and 5.12b.

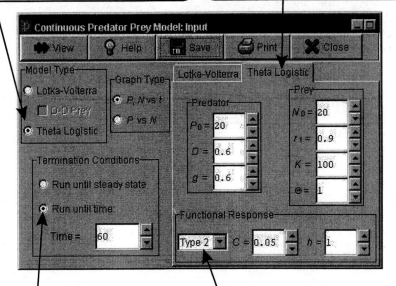

The default termination option for this module is a fixed runtime, because many cases never reach stable equilibrium densities.

The first window in the functional response box offers Type I, II, or III functional responses of equation (5.22), (5.23), or (5.24), respectively.

smaller proportion of the available prey as it becomes more saturated. This causes a positive feedback of prey density on prey-population growth because more and more predators are required to hold the prey in check as prey-population density increases. Negative, intraspecific feedback also increases with prey density and has the opposite influence, reducing the number of predators that the prey population can support at equilibrium. This effect becomes stronger as N approaches K, giving the isocline a negative slope on the right side of its hump. At $N = K$ and $P = 0$, prey density is entirely regulated by intraspecific density-dependent feedback.

The corresponding predator isocline is found by substituting the Type II function response [equation (5.23)] into equation (5.21), setting the result equal to zero and solving for N.

$$\frac{dP}{dt} = gP\left[\frac{CN}{1 + hCN} - D\right] \tag{5.27}$$

$$N = \frac{D}{C(1 - Dh)} \tag{5.28}$$

Because the ratio of constants is also constant, this θ-logistic predator isocline is a straight vertical line on the N–P plane just as it was for the Lotka-Volterra model.

Figures 5.11 and 5.12 compare the dynamics of θ-logistic simulations with Type I and Type II functional responses. Parameter values are the same in both simulations. The Type I functional response on the left produces a stable, damped oscillation. With the saturating Type II functional response on the right, density fluctuations increase from a start near the equilibrium point, then settle into a sustained oscillation of constant period and amplitude called a **limit cycle**. This cycle is different from the neutrally

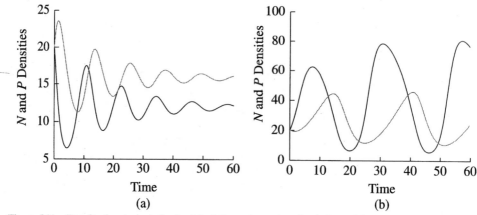

Figure 5.11   *Populus* time trajectories for θ-logistic predator-prey simulations with Type I and Type II functional responses. Parameter values for both runs were $N_0 = 20, r_1 = 0.9, K = 100, = 1, P_0 = 20, D = 0.6, g = 0.6$, and $C = 0.05$. Case (a) on the left models a linear, Type I functional response. Intraspecific, density-dependent feedback damps the oscillation to a stable equilibrium. In case (b) on the right, a Type II response (with $h = 1$) causes the oscillation to expand to a limit cycle of constant period and amplitude. The same two simulations are illustrated with isoclines on an N–P plane in Figure 5.12.

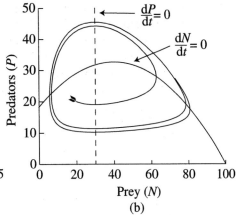

Figure 5.12   *Populus* phase-plane trajectories for the θ-logistic predator-prey simulations illustrated in Figure 5.11. The humped prey isocline in the Type II simulation reflects increases in both predator saturation, and intraspecific, density-dependent feedback among the prey. The saturation effect predominates at low density, while the intraspecific density dependence predominates at high prey density.

stable oscillation of a Lotka-Volterra model because the system returns to a persistent oscillation of the same period and amplitude if it is perturbed to either lower or higher densities. If the run had begun with N_0 and P_0 values outside the limit cycle of Figure 5.12, it would have spiraled inward to the same cycle. This tells us that a saturating, Type II functional response is less stable than the proportional Type I version; it maintains oscillations under conditions that lead to a constant equilibrial density with the Type I response. The Type II simulation is more consistent with lynx-hare data because it produces a limit cycle of constant period and amplitude, independent of starting densities or environmental variation.

THE PARADOX OF ENRICHMENT

The isoclines in Figure 5.12 illustrate a general pattern. In the Type I simulation on the left, the prey isocline has a negative slope where it crosses the predator isocline, an arrangement that results in convergence to stable equilibrium densities (see also Figure 5.9). If the prey isocline crosses the predator isocline with a positive slope, the dynamics lead to a sustained limit cycle (Rosenzweig and MacArthur 1963).

The humped prey isocline of Figure 5.12b has both positive and negative slopes, so the overall stability of this interaction depends on where the isoclines cross. If the predator isocline crosses the prey hump on the right, where its slope is negative, we expect a stable, converging oscillation. If it crosses on the left side of the hump (where the prey isocline has a positive slope), we expect a sustained limit cycle. Rosenzweig (1969, 1972) observed that the shape and position of a humped prey isocline depend on habitat productivity. Increasing productivity raises the carrying capacity, shifting the entire prey hump to the right. A predator isocline that formerly crossed the downslope might now cross the upslope of the prey isocline, changing a converging oscillation to a sustained cycle. Figure 5.13 shows an example. Carrying capacity increases from $K = 100$ on the left to $K = 120$ on the right, converting the damped oscillation into a persistent cycle. It is counterintuitive that increasing habitat quality would make predator-prey dynamics less stable, and Rosenzweig calls this effect of productivity on stability the paradox of enrichment. A similarly counterintuitive effect occurs with the density-dependent Lotka-Volterra model of Figures 5.8 and 5.9. Prey carrying capacity in these simulations is reduced from $K = 100$ (left) to $K = 30$ (right). Both simulations converge to stable equilibria, but it is the equilibrial density of predators, not prey, that declines with the reduction in prey carrying capacity.

POSTSCRIPT: WHAT ABOUT θ?

The partial implementation of equations (5.20) and (5.21) lacks one significant feature of the full θ-logistic predator-prey model as it was originally proposed by Gilpin and Ayala (1973). Their intraspecific prey-population-growth term, $rN(1 - (N/K)^\theta)$, has an additional parameter, represented by the Greek letter theta, that allows nonlinear feedback of prey-population density on prey growth. The paired equations for prey- and predator-population growth are then

$$\frac{dN}{dt} = r_1 N \left(1 - \left(\frac{N}{K} \right)^\theta \right) - f(N)P \tag{5.29}$$

$$\frac{dP}{dt} = gP[f(N) - D] \tag{5.30}$$

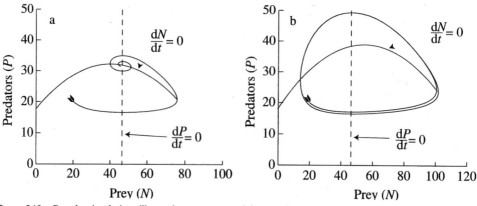

Figure 5.13   *Populus* simulations illustrating one aspect of the paradox of enrichment. The example on the left uses the same parameter values as Figure 5.12b ($N_0 = P_0 = 20, r_1 = 0.9, K = 100, = 1, g = 0.6, C = 0.05$) except that the predators' breakeven prey requirement is increased from $D = 0.6$ to $D = 0.7$. This moves the predator isocline from $N = 30$ to $N = 46.6$, so that it now crosses a falling section of the humped prey isocline. This changes the limit cycle of Figure 5.12b into a converging oscillation. On the right, the prey carrying capacity has been increased to $K = 130$, stretching the humped prey isocline to the right. Once again the predator isocline crosses an ascending portion of the hump, yielding a sustained limit cycle.

Table 5.1

SUMMARY OF EQUATIONS USED IN THE PREDATOR-PREY MODELS OF CHAPTER 5.

Model	Prey Equation	Predator Equation
Lotka-Volterra	$$\frac{dN}{dt} = r_1N - CNP$$	$$\frac{dP}{dt} = -d_2P + gCNP$$
Lotka-Volterra with density-dependent prey	$$\frac{dN}{dt} = r_1N\left(1 - \frac{N}{K}\right) - CNP$$	$$\frac{dP}{dt} = -d_2P + gCNP$$
θ-Logistic (partial implementation)	$$\frac{dN}{dt} = r_1N\left(1 - \frac{N}{K}\right) - f(N)P$$	$$\frac{dP}{dt} = gP[f(N) - D]$$
θ-Logistic	$$\frac{dN}{dt} = r_1N\left(1 - \left(\frac{N}{K}\right)^\theta\right) - f(N)P$$	$$\frac{dP}{dt} = gP[f(N) - D]$$

If θ is large ($\theta \gg 1$), the probabilities of birth and death change little until the prey population approaches its carrying capacity. If θ is small ($\theta \ll 1$), *per capita* birth or death rates (or both) decrease rapidly with increasing population size, even at low densities. When $\theta = 1$, per capita growth falls as a linear function of population size so that the intraspecific growth term in equation (5.29) is logistic. We will leave consideration of nonlinear density dependence and the dynamic implications of a Type III functional response to advanced students in the problems and exercises that follow.

Table 5.1 lists the models developed in this chapter. There are many potential influences on predator-prey interaction that we have not considered. For example, (a) foraging predators may interfere with one another so that the functional response becomes a function of both prey and predator densities; (b) polyphagous predators (those that eat several different kinds of prey) will be less tightly coupled to individual prey species than the simple pair-wise interactions modeled here; (c) discrete predator-prey models appropriate for organisms with seasonal life cycles will have additional properties associated with time lags and the finite-difference equations used to formulate them. Finally, (d) populations may be linked on the landscape mosaic to varying degree by dispersal of predators or prey, influencing local abundance. References that follow provide entry into the diverse literature addressing these issues.

REFERENCES

Case, T. J., *An Illustrated Guide to Theoretical Ecology*. New York: Oxford University Press, 2000.

Elton, C, "The Canadian Snowshoe Rabbit Enquiry," *Canad. Field-Nat*, 1933, 47:63–69, 84–86.

Elton, C. and M. Nicholson, "The Ten-Year Cycle in Numbers of the Lynx in Canada," *J. Anim. Ecol.*, 1942, 11:215–244.

Gilpin, M. E. and F. J. Ayala, "Global Models of Growth and Competition," *Proc. Nat. Acad. Sci.*, 1973, 70:3590–3593.

Gotelli, N. J. , *A Primer of Ecology*. Sunderland, MA: Sinauer Associates, 1995.

Hassell, M. P., *The Dynamics of Arthropod Predator-Prey Systems*. Princeton: Princeton University Press, 1978.

Hewitt, C. G., *The Conservation of the Wild Life of Canada*. New York: C. Scribner's Sons, 1921.

Holling, C. S., "The Components of Predation as Revealed by a Study of Small Mammal Predation of the European Pine Sawfly," *The Canadian Entomologist*, 1959, 91:293–320.

Holling, C. S., "The Functional Response of Predators to Prey Density and its Role in Mimicry and Population Regulation," *Mem. Ent. Soc. Can.* #45, 1965.

Holling, C. S., "The Functional Response of Invertebrate Predators to Prey Density, *Mem. Ent. Soc. Can.* #48.

Lotka, A. J., *Elements of Physical Biology*. Baltimore: Williams & Wilkins, 1925. Reissued as *Elements of Mathematical Biology*. New York: Dover Publications Inc., 1956.

MacLulich, D. A., *Fluctuations in the Numbers of the Varying Hare (Lepus americanus)*, University of Toronto Studies, Biological Series No. 43. The University of Toronto Press, 1937.

May, R. M., *Stability and Complexity in Model Ecosystems*. Princeton: Princeton Universtiy Press, 1975.

Murdoch, W. M., "Switching in General Predators: Experiments on Predator Specificity and Stability of prey Populations," *Ecological Monographs*, 1969, 39:335–354.

Poland, H., *Fur-Bearing Animals in Nature and in Commerce*. London: Gurney & Jackson, 1892.

Rosenzweig, M.L., "Why the Prey Curve Has a Hump," *Amer. Nat.*, 1969, 103:81–87.

Rosenzweig, M. L., "Stability of Enriched Aquatic Ecosystems," *Science*, 1972, 175:564–565.

Rosenzweig, M. L. and R. H. MacArthur, "Graphical Representation and Stability Conditions of Predator-Prey Interactions," *American Naturalist*, 1963, 97:209–223.

Seton, E. T., *The Arctic Prairies; A Canoe-Journey of 2,000 Miles in Search of the Caribou; Being the Account of a Voyage to the Region North of Aylmer Lake*. New York: C. Scribner's Sons, 1911.

Volterra, V., "Fluctuations in the Abundance of a Species Considered Mathematically," *Nature*, 1926, 118:558–560.

PROBLEMS AND EXERCISES

LOTKA-VOLTERRA PREDATORS AND PREY

1. Here are the Lotka-Volterra predator-prey equations reformatted with parentheses to divide each equation into three separate terms:

$$\left(\frac{dN}{dt}\right) = (r_1 N) - (CNP)$$

$$\left(\frac{dP}{dt}\right) = (d_2 N) - (gCNP)$$

Use your own words to define the component parameters and explain the significance of each term within parentheses.

2. What dynamics would you expect from this Lotka-Volterra predator-prey model if $P_0 = 0$? If $N_0 = 0$? Use the equations above to explain your prediction; then check it with *Populus* simulations.

3. Notice that the oscillating abundance of predators and prey in Figure 5.2a forms a closed loop which is repeatedly retraced in Figure 5.2b. Do the dynam-

ics move clockwise or counterclockwise around this loop? By choosing alter-nate starting densities so that the interaction begins in different quadrants of Figure 5.2b, can you reverse the direction of progress around the loop? Why? Formulate your answer from the equations or from the isocline graphs of Figures 5.3–5.5, then test it with *Populus* simulations.

4. In the Lotka-Volterra isocline diagram of Figure 5.5, the $dP/dt = 0$ isocline is vertical, striking only the N axis, and the $dN/dt = 0$ isocline is horizontal, striking only the P axis. What interaction between the two species would be implied if the isoclines switched place, so that the $dP/dt = 0$ isocline was hori-zontal, striking only the P axis and the $dN/dt = 0$ axis was vertical, striking only the N axis?

5. Explain what Lotka-Volterra assumptions make it possible for large and small prey population densities to be held at zero net growth ($dN/dt = 0$) by the same constant number of predators.

6. Run the *Populus* simulation of Lotka-Volterra predators and prey with the fol-lowing parameter values: $N_0 = P_0 = 20$, $r_1 = 0.1$, $d_2 = 0.1$, $C = 0.01$, and $g = 1$. Examine both the time-trajectory and phase-plane outputs of the simulation. What are the equilibrial densities of predators and prey? Initially, the predators increase and the prey decrease. Why does this occur? How can you change either N_0 or P_0 so that predator density will decrease initially? How can you change either N_0 or P_0 so that prey density will increase initially? Explain the effects of these changes, referring to the isocline equations, (5.7) and (5.11), and the isocline graph of Figure 5.5.

7. Run a new *Populus* simulation using the same values, $N_0 = P_0 = 20$, $C = 0.01$, and $g = 1$, but this time double the rate constants, setting $r_1 = d_2 = 0.2$. Examine both output screens, and explain the result of this simulation. If you need a hint, run another simulation, this time changing the rate constants to $r_1 = d_2 = 0.201$.

8. Reset the *Populus* parameter window to $N_0 = P_0 = 20$, $r_1 = d_2 = 0.1$, $C_1 = 0.01$, and $g = 1$. What happens if you triple the prey growth rate to $r_1 = 0.3$, holding all other parameter values at their defaults? Refer to the equations and iso-clines to formulate an *a priori* answer for this question; then use *Populus* simu-lation to test your prediction. Your predictions and tests should address changes in both the equilibrial densities and the amplitude of oscillations caused by this increase in the prey growth rate. What would have happened if instead of tripling the prey growth rate, r_1, you had tripled the predator starva-tion rate, d_2?

9. The Lotka-Volterra model assumes that both prey and predator growth are density independent. The addition of negative density-dependent feedback to the prey equation changed the flat prey isocline of Figure 5.7 to the negatively sloped isoclines in Figure 5.9. What effect would direct negative density-depen-dent feedback among predators have on the shape of the predator isocline?

θ-LOGISTIC PREDATOR-PREY MODELS

10. Use your own words to explain the difference between a neutrally stable oscillation and a limit cycle.

11. Write out the prey and predator equations for a θ-logistic model with a Type I functional response by substituting equation (5.22) into equations (5.20) and (5.21). Now set these equations equal to zero and simplify to derive predator- and prey-isocline equations analogous to (5.26) and (5.28). Set *Populus* parameter values for a Type I θ-logistic to $N_0 = 10, r_1 = 0.9, K = 100$, and $θ = 1$ for the prey, and $P_0 = 10, D = 1, g = 0.5$, and $C = 0.05$ for the predators. Using these values in your isocline equations, determine the y-intercept of the prey isocline graphed on an $N–P$ plane, the x-intercepts of both predator and prey isoclines, and the joint equilibrial densities where the isoclines cross. Finally, check your efforts with a *Populus* simulation.

12. If environmental productivity increases, raising the carrying capacity, K, of the prey, how would you expect equilibrial densities of θ-logistic prey and predators with a Type I functional response to change? Explain your *a priori* prediction; then check your intuition with a *Populus* simulation, doubling the prey carrying capacity from the preceding problem to $K = 200$.

13. *Populus* default parameter values for the θ-logistic model with a Type II functional response are the same as those given in Problem 11 with one additional parameter, the handling time, $h = 1$. Run this default simulation and examine both the time-trajectory and phase-plane output graphs. What parameter value(s) would you change to simulate Rosenzweig's paradox of enrichment? Check your answer with a simulation, and explain the results using your own words.

14. Repeat Problem 11, this time substituting a Type III functional response [equation (5.24)] into the θ-logistic model, and completing the same analyses. When you have finished, use your own words to compare and explain the dynamic behaviors and relative stability conferred by the three functional-response patterns.

FOR ADVANCED STUDENTS

15. The $θ$ in a θ-logistic prey equation allows density-dependent feedback of prey-population size on prey-population growth to be nonlinear. With $θ > 1$, the prey isocline has a less negative slope until N approaches K, then drops precipitously. Predict what effect this will have on the equilibrial densities and the time required for oscillations to damp out for a θ-logistic simulation with a Type I functional response. Test your prediction with *Populus* simulations; leave all parameters at their defaults, but try $θ = 1, θ = 2, θ = 5$. Make analogous predictions about the effect of $θ < 1, θ = 0$, and test your expectations, running $θ = 1, θ = 0.5, θ = 0.1, θ = 0$.

CHAPTER 6
Infectious Microparasitic Diseases

Parasites ranging from viruses and bacteria to fungi, helminths, and arthropods can cause infectious diseases that have a strong influence on the dynamics of their hosts. Roughly a third of the inhabitants of Europe died from plague during the Black Death pandemic of 1348 and 1349 (Gasquet 1908, Ziegler 1969). The World Health Organization (WHO) estimates that 33.6 million people worldwide carried the human immunodeficiency virus (AIDS) at the end of 1999. Most of those people will die within the next decade, and 70% of this mortality will be concentrated in sub-Saharan Africa, which has already sustained 84% of the 16.3 million worldwide AIDS deaths since the beginning of the epidemic (*http://www.who.int/emc-hiv/una99e53,doc*).

Parasites can also serve our management objectives, offering a biological means for controlling agricultural pests. From 1950 to 1952 a *Myxoma* pox virus of the tropical forest rabbit, *Sylvilagus brasiliensis*, was introduced into Australia to control European rabbits, *Oryctolagus cuniculus*. The resulting myxomatosis produced case-mortality rates of 99.8%, rabbit populations fell precipitously, and survivors were subsequently restricted to marginal, nonagricultural habitats (Fenner and Myers 1978).

Host-parasite interactions differ from those of predators and prey because the diseases do not necessarily kill the host and because recovered hosts may develop a longterm immunity to re-infection. To simplify the analysis of disease dynamics, it is useful to distinguish between micro- and macroparasites. Microparasites (like the *Variola* viruses that cause smallpox) reproduce quickly, reaching tremendous population densities within a host. The duration of these acute infections is often limited by host defenses and short relative to host lifespan, and recovered victims may develop lifetime immunity. Microparasite dynamics are driven largely by transmission between hosts; their essentials can be captured by models that classify host individuals as susceptible, infected, or recovered (immune), without accounting for within-host abundance of the parasite. In contrast, macroparasites (like intestinal flatworms) typically cause chronic and persistent infections. Disease severity depends on the number of parasites present, and a small fraction of the host population may harbor most of the parasites. Satisfactory models of macroparasite dynamics (*cf.* May and Anderson 1979, Dobson and Hudson 1992) must account for parasite abundance within individual hosts and host-to-host variation in parasite abundance. The natural history of some plant diseases requires a similarly detailed approach. As a result, models of macroparasites and plant diseases are more complex than the microparasite models that will be explored in this introduction.

The *Populus* microparasite simulation presents a model developed by Anderson and May (1979, 1982) from classic approaches of Ross (1916, 1917) and Kermack and McKendrick (1927) and includes an alternative transmission process detailed by Getz and Pickering (1983). The model describes a host population of size N, containing susceptible individuals (S) who are not infected, and infected individuals (I) who can pass the parasite to others. It may also include recovered individuals (R) who have encountered the disease and developed immunity. Susceptible individuals arise through birth or the loss of immunity at *per capita* rates b and γ, respectively. Individuals leave the susceptible class through natural mortality (rate d) or by acquiring the parasite (rate β) after encountering an infected host. Individuals leave the infected category through natural mortality (rate d), disease-induced mortality (rate α), or through recovery (rate υ) of hosts that become immune.

The model assumes that (1) individuals are uninfected at birth, (2) newly infected hosts can transmit the disease immediately, (3) there is no age structure among hosts, (4) the disease does not affect host fecundity, and (5) host populations are large enough so that random events can be ignored. The basic Anderson and May model assumes further that (6) there is no density-dependent feedback among hosts other than parasite-induced mortality and (7) infections occur in direct proportion to the number of encounters between susceptible and infected individuals, set by the product of their respective population densities (SI). We call this **mass-action** infection process **density-dependent transmission**.

To summarize, the model parameters are:

N = total host population density	d = host natural mortality rate
S = susceptible host density	α = disease-induced mortality rate
I = infected host density	β = between-host transmission rate
R = recovered (immune) host density	υ = recovery rate of infected hosts
b = host birth rate	γ = rate of immunity loss

6.1 *SI* MODEL WITH DENSITY-DEPENDENT TRANSMISSION

Based on these assumptions, we can specify a differential equation for the dynamics of each host class. We will begin with a version of the model illustrated schematically in Figure 6.1. This *SI* model divides the host population into just two groups, susceptible and infected.

$$\frac{dS}{dt} = b(S + I) + \nu I - dS - \beta SI \tag{6.1}$$

$$\frac{dI}{dt} = \beta SI - (\alpha + d + \nu)I \tag{6.2}$$

The terms in each equation represent transition arrows in Figure 6.1. For example, the two positive terms in equation (6.1) represent increments to the susceptible (S) class resulting from birth and recovery of infected individuals. The two negative terms represent losses due to natural mortality and disease transmission. The transmission term, $-\beta SI$, includes the encounter rate between susceptible and infected hosts, given by the product of their densities, SI, and the probability of disease transmission per encounter, β. The same transmission term contributes a positive increment to the growth of infected hosts [equation (6.2)]. The negative term summarizes three differ-

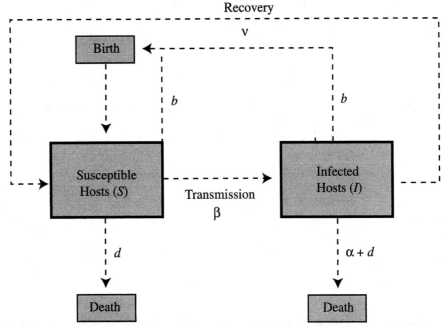

Figure 6.1 Schematic of a *SI* host-microparasite model, with the host population divided into susceptible and infected classes, simplified and redrawn after Anderson and May 1982. This version is appropriate in some arthropod host-parasite systems that lack acquired immunity; for systems with an immune response, it is still a convenient simplification because it allows graphical analyses on a two-dimensional phase plane.

ent rate processes that remove infected hosts from the population: disease-induced mortality, α; natural mortality, d; and recovery, v. We can add the expressions for the two host classes to specify change in total host population size,

$$\frac{dN}{dt} = (b - d)N - \alpha I \tag{6.3}$$

6.2 DYNAMICS AND EQUILIBRIA

Although microparasites often reproduce rapidly within a susceptible host, transmission between hosts is the critical feature of their dynamics. We define the **net reproductive rate** of a disease as the number of newly infected individuals produced by a single infected host introduced into a population of susceptible hosts. Dividing through equation (6.2) by I, the *per capita* increase of infected individuals is

$$\frac{dI}{I dt} = \beta S - (\alpha + d + v) \tag{6.4}$$

This implies that the infected class is incremented by a product of transmission rate (β) and the number of susceptibles (S), and decremented by disease-induced mortality (α), natural mortality (d), and host recovery (v). These three rate parameters, α, d and

v, have units of $1/t$. The reciprocal of their sum, $1/(\alpha + d + v)$, is the average interval during which an infected individual can transmit the disease. The number of new cases that it will cause during this interval is βS. So the net reproductive rate, R_0, is

$$R_0 = \frac{\beta S}{\alpha + d + v} \tag{6.5}$$

Intuitively, it might seem reasonable to derive this net reproductive rate by integrating equation (6..2) over the average persistence of infection, but such an estimate would include cases caused by secondary transmission. R_0 is the number of new cases acquired directly from a single infected individual. Notice that net reproduction increases in direct proportion to susceptible host density, S.

R_0 must equal or exceed 1.0 for a disease to persist or spread in its host population because each infected individual must pass the parasite to at least 1 uninfected host before dying or recovering. We can set $R_0 = 1$ in equation (6.5) and rearrange to find the minimum susceptible host density threshold (S_T) necessary to sustain the parasite (assuming that the various rates remain constant and the population is homogeneously mixed).

$$S_T = \frac{\alpha + d + v}{\beta} \tag{6.6}$$

S_T, the disease's threshold susceptible-host density, is a minimum number of susceptible hosts required by the parasite, which will decline to extinction unless $S \geq S_T$. When the population density of susceptible hosts, S, is above S_T, then $R_0 > 1$ and we can rewrite equation (6.5), substituting $1/S_T$ for $\beta/\alpha + d + v$ [from equation (6.6)] to give

$$R_0 = \frac{S}{S_T} \tag{6.7}$$

Because the net reproduction of a disease increases with susceptible host density, it is interesting to ask whether microparasitic diseases might regulate their host populations. If we define the **intrinsic growth rate** of uninfected hosts as $r = b - d$ and the **prevalence** of a disease (the infected fraction of the host population) as $y = I/N$, then equation (6.3) for change in total host population density can be rearranged to give

$$\frac{dN}{dt} = (r - \alpha y)N \tag{6.8}$$

This shows that a sufficient prevalence (y) and disease-induced mortality rate (α) will balance intrinsic growth (r), and the disease can regulate host density at a stable value, N. The prevalence of the disease at this equilibrium is found by setting $dN/dt = 0$ in equation (6.8), giving

$$0 = (r - \alpha y)N \tag{6.9}$$

$$\hat{y} = \frac{r}{\alpha} \tag{6.10}$$

We can analyze the dynamics of this *SI* model graphically by plotting isoclines on a phase plane. Because the total host population size (N) is the sum of its susceptible (S) and infected (I) classes, either an S vs I or an N vs I phase plane could be used; the N vs I graph is especially valuable because it shows the effect of parasites on host-population stability and equilibrial density. Isoclines for an N vs I phase plane are found by setting equations (6.3) and (6.2) equal to zero and rearranging. First, rearranging equation (6.3) to find the N isocline,

$$\frac{dN}{dt} = 0 = (b - d)N - \alpha I \tag{6.11}$$

$$I = \frac{(b - d)}{\alpha}N = \frac{r}{\alpha}N \tag{6.12}$$

This is a straight line with a y-intercept at $N = 0$ and a slope of r/α (Figure 6.2). Above the N isocline, the density of infected hosts and disease-induced mortality cause total host-population density to decline.

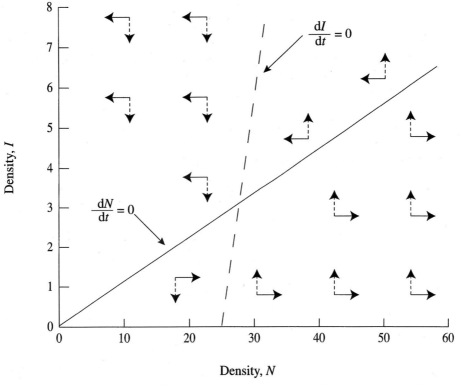

Figure 6.2   Isoclines for an *SI* model with density-dependent parasite transmission, drawn on the N vs I phase plane. Total host population size, N, is the sum of susceptible and infected host classes, $S + I$. Therefore the point where the N and I isoclines cross marks an equilibrium. Growth vectors in the four quadrants around the equilibrium show that the model cycles, and *Populus* simulations (Figure 6.3) demonstrate that the equilibrium is stable. With this isocline configuration a parasite with density-dependent transmission regulates its host.

Setting equation (6.2) equal to 0 and rearranging to find the corresponding I isocline, we have

$$\frac{dI}{dt} = 0 = \beta SI - (\alpha + d + v)I \tag{6.13}$$

$$I = N - \left(\frac{\alpha + d + v}{\beta}\right) \tag{6.14}$$

The I isocline is a line with a slope of 1 and a y-intercept of $-(\alpha + d + v)/\beta$, as shown in Figure 6.2. To the left of this line, susceptible host density is too low to sustain $dI/dt = 0$, and the number of infected hosts declines. To the right of the isocline, infected-host density increases.

To find the densities of susceptible and infected hosts at the equilibrium where the isoclines cross and disease-induced mortality balances the intrinsic growth capacity of the host population, we begin with the net reproductive rate. At equilibrium, each infected individual will replace itself on average with a single new case, so susceptible hosts must be at the threshold density given by equation (6.6),

$$\hat{S}_T = \frac{\alpha + d + v}{\beta} \tag{6.15}$$

Because the total host population size, N, is the sum of susceptible and infected hosts in the SI model, we can now substitute $(\hat{N} - \hat{S})$ for I in the N isocline equation (6.12), and rearrange to solve for $\hat{N}$:

$$N - S = \frac{r}{\alpha}N \tag{6.16}$$

Substituting equation (6.15) for S, we have

$$N - \left(\frac{\alpha + d + v}{\beta}\right) = \frac{r}{\alpha}N \tag{6.17}$$

$$N\left(1 - \frac{r}{\alpha}\right) = \left(\frac{a + d + v}{\beta}\right) \tag{6.18}$$

$$\hat{N} = \frac{\alpha}{\alpha + r}\left(\frac{\alpha + d + v}{\beta}\right) \tag{6.19}$$

To determine the corresponding equilibrial density of infected hosts, we calculate $\hat{I}$ as $\hat{N} - \hat{S}$.

$$\hat{I} = \frac{\alpha}{\alpha - r}\left(\frac{\alpha + d + v}{\beta}\right) - \left(\frac{a + d + v}{\beta}\right) \tag{6.20}$$

$$\hat{I} = \frac{\alpha}{\alpha - r}\left(\frac{\alpha + d + v}{\beta}\right) - \frac{a - r}{\alpha - r}\left(\frac{\alpha + d + v}{\beta}\right) \tag{6.21}$$

$$\hat{I} = \frac{\alpha}{\alpha - r}\left(\frac{\alpha + d + \nu}{\beta}\right) \tag{6.22}$$

In addition to these analytical characterizations of the *SI* model, we can also use numerical integration to simulate its dynamics with a computer. Figure 6.3 illustrates a *Populus SI* simulation with density-dependent transmission, based on the parameter values and option settings of Box 6.1. The time trajectory on the left shows a damped oscillation of susceptible, infected, and total host populations as the disease regulates

BOX 6.1

The *Populus* input window for a *SI* host-microparasite simulation with density-dependent transmission. These parameter values and option settings produce the damped oscillation of Figure 6.3a. Selecting the *N* vs *I* plot-type option gives the phase-plane output graph shown in Figure 6.3b.

This simulation divides the host population into two or three classes. When the *SI* version is selected, the *R*-host density window is closed. Each value can be toggled up or down and the coupled output changes immediately.

Host rate parameters are represented by arabic symbols, and parasite parameters are greek. The loss-of-immunity parameter γ is not used for *SI* versions; the window is closed and gray.

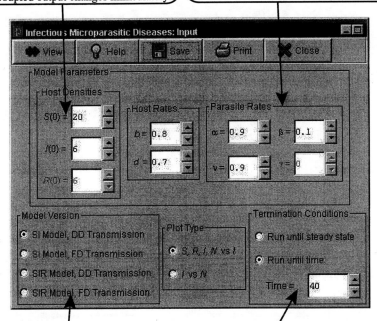

Density-dependent versions use the mass-action transmission term β*SI*. The frequency-dependent analog is β*SI/N*.

By default, the simulation terminates after a fixed run time because frequency-dependent runs are unregulated.

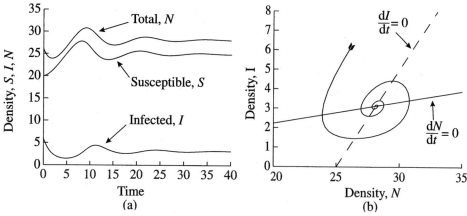

Figure 6.3 *Populus* simulation of a host-microparasite interaction using the *SI* model with density-dependent transmission. Parameter values and option settings that produce this output are illustrated in Box 6.1. Graph (a) shows the time trajectory of *S*, *I*, and *N* densities; graph (b) shows the same damped oscillation on an *N* vs *I* phase plane with the *N* and *I* isoclines.

host density at a stable equilibrium. The *N* vs *I* phase-plane plot on the right shows the same simulation as a spiral leading inward to the equilibrium where the isoclines cross.

6.3 *SI* MODEL WITH FREQUENCY-DEPENDENT TRANSMISSION

The homogeneous mixing and mass-action mating assumed in equations (6.1) and (6.2) (particularly in the density-dependent transmission term, βSI) are often inappropriate. For example, in many species individuals have a limited number of sexual partners. A sexually transmitted disease will spread in proportion to its frequency among mates, and the number of diseased individuals in the population as a whole is only relevant if it sets disease frequency in the subset of partners. Getz and Pickering (1983) add frequency-dependent transmission to the basic system of equations (7.1) and (7.2), so that

$$\frac{dS}{dt} = b(S + I) + \nu I - dS - \frac{\beta SI}{N} \tag{6.23}$$

$$\frac{dI}{dt} = \frac{\beta SI}{N} - (\alpha + d + \nu)I \tag{6.24}$$

Now the transmission term has been changed from the simple product βSI, where the number of new cases is proportional to the absolute numbers of susceptible hosts, to $\beta SI/N$. Here the frequency of susceptible hosts, S/N, determines the number of transmissions rather than the absolute density, S. These two equations (6.23) and (6.24) can be summed to produce a differential equation for total population growth identical to equation (6.3).

6.4 *SI* DYNAMICS WITH FREQUENCY-DEPENDENT TRANSMISSION

The dynamics of this system are fundamentally different from those of the *SI* model with density-dependent transmission. The net reproductive rate, R_0 is proportional not to susceptible-host density as it was in equation (6.5), but to susceptible-host frequency, S/N.

$$R_0 = \frac{\beta S}{(\alpha + d + \nu)N} \tag{6.25}$$

When a disease is introduced into a previously uninfected population, its initial prevalence is very low so $S \approx N$. This means that $S/N \approx 1$, and therefore

$$R_0 \approx \frac{\beta}{(\alpha + d + \nu)} \tag{6.26}$$

If this ratio of constants exceeds $R_0 = 1$, the disease can invade and spread. Thus a parasite with frequency-dependent transmission has no threshold susceptible-host density. Sexually transmitted diseases can invade and sustain themselves even in very small host populations. Their only essential requirement is that hosts must be sufficiently promiscuous to maintain the chain of infection. Parasites must be acquired from one partner and passed to another.

Because the differential equation (6.3) projecting changes in total host-population size is the same for frequency- and density-dependent *SI* models, the N isocline remains as it was in equation (6.12). On the N vs I phase plane, this is a straight line with a y-intercept at $N = 0$ and a slope of r/α. The I isocline for an *SI* model with frequency-dependent transmission is found by setting equation (6.24) equal to zero and solving for I as a function of N.

$$\frac{dI}{dt} = 0 = \frac{\beta SI}{N} - (\alpha + d + \nu)I \tag{6.27}$$

$$0 = \frac{\beta(N - I)}{N} - (\alpha + d + \nu) \tag{6.28}$$

$$I = \beta N - N(\alpha + d + \nu) \tag{6.29}$$

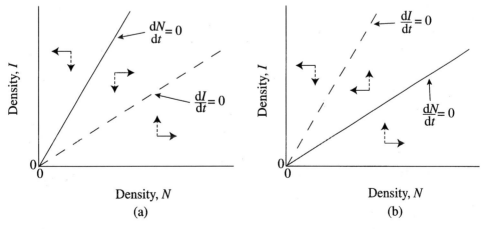

(a)   **(b)**

Figure 6.4   Schematic representation of the two possible isocline arrangements for an *SI* model with frequency-dependent parasite transmission. In the example on the left, parasites are not able to compensate for the population growth of hosts, which increase exponentially. In the example on the right, host birth lags behind the sum of natural and disease-induced mortality, leading both hosts and parasites to extinction.

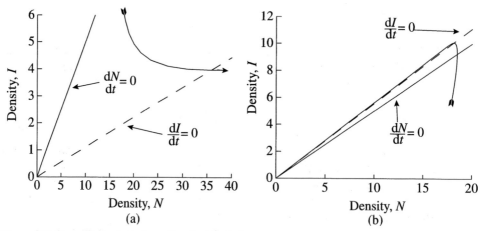

Figure 6.5 Phase-plane plots of two *Populus* simulations representing the corresponding left and right examples in Figure 6.4. Parameter values for simulation (a) on the left were α = 0.4, β = 0.9, υ = 0.2, b = 0.4, d = 0.2, S(0) = 20, and I(0) = 1, resulting in exponential growth of the host population. Parameter values for simulation (b) on the right were α = 0.1, β = 0.9, υ = 0.1, b = 0.25, d = 0.2, S(0) = 20, and I(0) = 1, resulting in the extinction of both hosts and parasites.

$$I = N - N\left(\frac{\alpha + d + v}{\beta}\right) \tag{6.30}$$

This is a line with a slope < 1 and a y-intercept at $N = 0$. Since the N and I isoclines cross at $N = I = 0$, there is no equilibrium in the positive quadrant of the phase plane, where $N > 0 < I$. This means that a parasite with frequency-dependent transmission cannot regulate its host. There are two possible arrangements of the I and N isoclines, illustrated with growth vectors in Figure 6.4. If the N isocline has the steeper slope, disease-induced mortality is not sufficient to balance the reproduc-

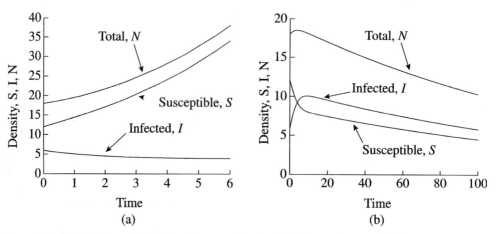

Figure 6.6 *S, I* and *N* time trajectories of the same *Populus* simulations represented in the corresponding left and right phase planes of Figure 6.5.

tion of hosts, which grow exponentially. If the *I* isocline has the steeper slope, both hosts and parasites decline toward extinction. Figures 6.5 and 6.6 show phase planes and time trajectories, respectively, of *Populus* simulations that illustrate both outcomes.

6.5 *SIR* MODEL WITH DENSITY-DEPENDENT TRANSMISSION

The *SI* model of the preceding sections is a convenient simplification. With a system of two differential equations, we were able to use the analytical and graphical techniques developed in earlier chapters for paired interactions involving competition and predation. To track the dynamics of resistant hosts that have encountered the disease, recovered, and developed immunity, we need a third equation. There will also be additional terms in the *S* and *I* equations reflecting transitions to and from the resistant class. The analysis of a three-equation system is more complex, but in this case, the results are quite similar. In an *SIR* model, parasites with density-dependent transmission require a minimum threshold population of susceptible hosts, S_T, to invade and persist. Appropriate rate parameters allow them to regulate the host population at this equilibrial density. With the frequency-dependent transmission characteristic of sexually transmitted diseases, there is no threshold susceptible-host density required to maintain the parasite. Similarly, the host population is not regulated in an *SIR* model with frequency-dependent transmission unless we add some additional source of density-dependent feedback. We will cite these similarities to justify passing over a full analysis of *SIR* models, focusing only on features required to illustrate an important practical issue, the dynamics of immunization.

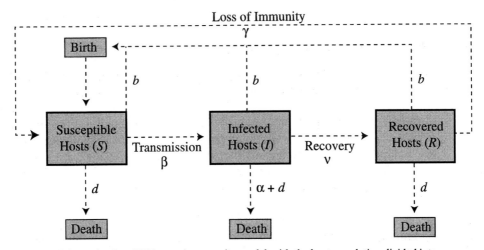

Figure 6.7   Schematic of an *SIR* host-microparasite model, with the host population divided into susceptible, infected, and resistant classes, redrawn after Anderson and May (1982). The recovery parameter now quantifies entry into the resistant class, rather than to the susceptible class, as in Figure 6.1. In addition, there is a new parameter, γ, quantifying the loss of immunity, and reentry into the susceptible class.

The *SIR* schematic in Figure 6.7 includes a new box for the R class of resistant hosts. The recovery arrow, υ, now leads to this R box instead of cycling back to the susceptible box as it did in Figure 6.1. There is also a new arrow for the loss of immunity, quantified by γ, the transition rate from R to S. The equations that we use to model changes in the three host classes are directly analogous to equations (6.1) and (6.2).

$$\frac{dS}{dt} = b(S + I + R) - dS - \beta SI + \gamma R \tag{6.31}$$

$$\frac{dI}{dt} = \beta SI - (\alpha + d + \upsilon)I \tag{6.32}$$

$$\frac{dR}{dt} = \upsilon I - (d + \gamma)R \tag{6.33}$$

To make a corresponding *SIR* model with frequency-dependent parasite transmission, we would change the transmission term βSI in equations (6.31) and (6.32) to $\beta SI/N$. The sum of these three equations giving the rate of change in size for the full host population, dN/dt, remains the same as the previous summation in equation (6.3). The net reproductive rate of a disease with density-dependent transmission [equation (6.5)] is still

$$R_0 = \frac{\beta S}{\alpha + d + \upsilon} \tag{6.34}$$

so the threshold susceptible-host population density required to sustain the parasite remains as it was given in equation (6.6).

$$S_T = \frac{\alpha + d + \upsilon}{\beta} \tag{6.35}$$

6.6 DYNAMICS OF IMMUNIZATION

One goal of immunization is to reduce the number of susceptible individuals in a host population, consequently lowering the net reproductive rate, R_0, of a disease. Recall from equation (6.7) that $R_0 = S/S_T$. The net reproduction of a disease is directly proportional to susceptible-host density (the proportionality constant is $1/S_T$, or $\beta/(\alpha + d + \upsilon)$, so parasites with high β, low α, and low υ are highly transmissible, with low S_T and high R_0). This means that every fractional reduction in susceptible-host density causes a comparable reduction in the net reproductive rate. Suppose that we immunize a fraction, p, of the host population; the new R_0' is

$$R_0' = (1 - p)R_0 \tag{6.36}$$

By immunizing a sufficient fraction of the host population, we could reduce S to a density below the minimum threshold, S_T, and eradicate the disease. The p that will hold $R_0' = 1$ is

$$R_0' = 1 = (1 - p)R_0 \tag{6.37}$$

so to drive the disease to extinction we need:

$$p > 1 - \frac{1}{R_0} \tag{6.38}$$

Diseases with a high net reproductive rate and a low threshold requirement for susceptible hosts are more difficult to eradicate than less transmissible diseases; their eradication requires the immunization of a larger fraction of the host population. Table 6.1 gives examples of the relation between net reproduction and the immunization fraction required for eradication, from data compiled by May (1983). The table also illustrates an effect of net reproduction on the average age at which hosts acquire a disease. Diseases that are highly transmissible are likely to infect their hosts at an early age. If L is the average life expectancy and A is the average age at which individuals acquire the infection, then

$$A = \frac{L}{R_0 - 1} \tag{6.39}$$

For example, if the net reproductive rate of chicken pox is $R_0 = 9$ and average life expectancy is $L = 75$, then the average age of hosts is 9.4. Childhood diseases are caused by parasites with high rates of net reproduction.

Because there is a direct relation between the net reproductive rate of a disease and the average age of infection, immunization and other practices that reduce the rate of spread increase the age at which individuals acquire a disease. Sometimes the delayed acquisition associated with public-health measures has unfortunate consequences. *Rubella* is dangerous for women of childbearing age because it causes birth defects. The same immunization program that protects a fraction of the host population reduces transmission of the disease, so that unprotected victims acquire it at a later and more dangerous point in their lives. Similarly, poliomyelitis had an R_0 of about 50 before modern civil engineering; it was a childhood disease, and a youngster's

---

Table 6.1

IMMUNIZATION FRACTION REQUIRED FOR DISEASE ERADICATION

Net reproductive rate, R_0, of common infectious diseases, with the average age at which the disease is acquired.

Disease	Locale	R_0	A	$p(\%)$
Smallpox (before campaign)	Developing Countries	3–5	18–37	70–80
Poliomyelitis	Holland 1960	6	15	83
Rubella	England & Wales 1979	6	15	83
Chicken pox	United States 1913–21	9–10	8–9	90
Measles	England & Wales 1956–68	13	6	92
Whooping cough	England & Wales 1942–50	17	4	94

Estimated using equation (6.39) assuming a life span, L, of 75 years and the immunization percentage, p, required for eradication (May 1983)

immune system usually dealt with it successfully. Water treatment has reduced the R_0 of polio to about 6, delaying disease onset and radically increasing its severity and paralytic consequences. The recently announced vaccine for chicken pox seems likely to cause a similar delay in acquisition of the disease among unprotected adolescents and adults, with more debilitating effects.

THE ERADICATION OF SMALLPOX

The mummy of the Egyptian pharoh Ramses V, who died from an acute illness in 1157 B.C., shows what are probably smallpox blisters; this disease has wreaked havoc through much of recorded human history (Behbehani 1988). The first widespread attempt at control involved variolation, injection of the lymph or puss from one smallpox victim into others. If this practice had any efficacy, it might have been through partial attenuation of the *Variola* virus in the source case, or passage to new individuals at a time of year or physical condition with improved prospects of survival. Variolation was widespread in England and America in the eighteenth century. Jenner's innovation, called vaccination after the Latin root for cow, involved inoculation with a cowpox virus (*Vaccinia*) similar enough to engender cross-reactive immunity against *Variola*. Vaccination came into widespread usage around 1800. There was considerable opposition early; the principle problems had largely to do with the primitive state of virology. There were multiple strains of the cowpox virus of varying effectiveness, and vaccines were often contaminated with active smallpox virus.

In this century, smallpox has been recognized as two distinctive but related diseases, *Variola major* (the classic or Asian smallpox) and *Variola minor* (African smallpox). *Variola major* has a case fatality rate of 15–45%, while that of *Variola minor* is negligible. The WHO further recognizes half a dozen different virulence strains of *Variola major* with case fatality rates ranging from 10–100%. There is a clear relationship between the virulence of the strains and clinical severity of the disease.

Smallpox is less contagious than influenza or measles, with a net reproductive rate, R_0 of 3 to 5. Transmission requires close personal contact. Its relatively slow spread through a community makes it an ideal candidate for eradication. The WHO initiated a campaign to eradicate smallpox via mass vaccination in the 1970s. Equation (6.38) suggests that the disease can be eradicated by immunizing the proportion

$$p > 1 - \frac{1}{4} = .75 \qquad (6.40)$$

of individuals in populations where the disease is present. Unfortunately, this strategy did not work. A 1969 study among 23 million people in central Java showed that more than 95% of the population bore the scars of successful vaccination. Theoretically, this effort should have driven R_0 below 1.0, yet in that year there were 1700 smallpox cases among the 5% who had not been vaccinated. This suggests that the population was not homogeneously mixed, with those individuals who were most difficult to find and vaccinate clustered in remote locations where the required immunization fraction (p) was not realized. Success was ultimately achieved by surveillance and containment; field workers advertised to find cases, and vaccinated all potential contacts of the victim to break the chain of infection (Behbehani 1988).

The last natural case of smallpox caused by *Variola major* was that of Rahima Banu (Compassionate Lady), a 3-year-old girl from the village of Kuralia on Bhola Island, Bangladesh, in November 1975; happily, she survived. A medical photographer, Janet Parker, was less fortunate when the virus escaped containment at an English research laboratory in 1978. Her death precipitated the suicide of the laboratory director and debate over the wisdom of maintaining *Variola* stocks, even for research (Behbehani 1988). In the decade since that academic debate, vaccination has been discontinued, susceptible-host density has grown enormously, and smallpox stocks may have spread widely with the diaspora of the Soviet biological weapons program.

6.7 THE EVOLUTION OF VIRULENCE

The rapid reproduction and short generation time of many microparasites confers a potential for very rapid evolution. A common suggestion about the likely course of parasite evolution is that they should become avirulent. Justification for this expectation comes from the net reproductive rate, $R_0 = \beta S / (\alpha + d + \nu)$. Disease-induced mortality, α, is in the denominator, so parasites that do less damage will allow infected hosts to remain alive and infectious over a longer interval and achieve higher net reproduction (Ewald 1993). This assumes that disease transmission and the duration of infectiousness are independent of virulence. In fact, there is often a direct mechanistic linkage between the damage caused by parasites and their transmission. The damage that rhinoviruses inflict on our sinus membranes causes nasal discharge and the sneezing that distributes viral particles in an infectious aerosol, spreading common colds. The linkage of virulence and transmission imposes a tradeoff on parasite evolution. Models suggest that the evolutionary trend depends on the exact form of this tradeoff and the abundance of susceptible hosts. Relatively virulent parasites prosper when susceptible hosts are common, but they may be displaced by forms with intermediate virulence and lower threshold requirements when the disease has reduced susceptible hosts to equilibrial density (Lenski and May 1994).

Empirical evidence from the history of myxomatosis in Australian rabbits supports this view that parasites may evolve toward intermediate virulence. The *Myxoma* strains that were first introduced were deadly, and they quickly decimated the huge rabbit populations plaguing Australian agriculture. However, the disease is vectored by mosquitoes, which are absent during the austral winter. As host density declined, selection favored viral strains that required a lower threshold susceptible-host density and also left the host vital enough to survive over winter. The strains that were first introduced in 1950–51 allowed a mean host survival < 13 days. By 1963–64, the most common virulence strain isolated from the wild allowed a mean host survival of 23–28 days, and this strain has predominated subsequently. A small percentage of the Australian isolates produce case-mortality rates below 50%, so average host survival is very long, but these relatively avirulent forms have not succeeded in displacing the strains with intermediate virulence from their modal frequency (Fenner and Myers 1978).

While this chapter has concerned itself largely with the ecology of microparasitic diseases, the issue of virulence is just one of many important questions about their evolution. Since parasites are an important public health concern, this biology has considerable medical significance. The evolution of antibiotic resistance provides another example. The problem of resistant pathogens is increasing and may force us to reconsider many practices in medicine and animal husbandry from a more evolutionary perspective. Students who would like to pursue these issues will find a good entry and introduction to this literature in Ewald (1994).

REFERENCES

Anderson, R. M., "Transmission Dynamics and Control of infectious Disease Agents." In R. M. Anderson and R. M. May (eds), *Population Biology of Infectious Diseases* (Dahlem Conference Report). Springer-Verlag, 1982.

Anderson, R. M. and R. M. May, "Population Biology of Infectious Diseases: Part I," *Nature*, 1979, 280:361–367.

Anderson, R. M. and R. M. May, "Directly Transmitted Infectious Diseases: Control by Vaccination," *Science*, 1982, 215:1053–1060.

Anderson, R. M. and R. M. May, *Infectious Disease in Humans.* New York: Oxford University Press, 1991.

Behbehani, A. M., *The Smallpox Story.* Kansas City: University of Kansas Medical Center, 1988.

Dobson, A. P. and P. J. Hudson, "*Regulation and Stability of a Free-Living Host-Parasite System: Trichostrongylus tenuis in Red Grouse. II. Population Models,*" *J. Anim. Ecol.*, 1992, 61:487–498.

Ewald, P. W., "The Evolution of Virulence," *Scientific American*, April 1993, 86–93.

Ewald, P. W., *Evolution of Infectious Disease.* Oxford: Oxford University Press, 1994.

Fenner, F. and K. Myers, "*Myxoma* Virus and Myxomatosis in Retrospect: The First Quarter Century of a New Disease." In E. Kurstak and K. Maramorosch (eds), *Viruses and Environment.* Academic Press, 1978.

Gasquet, Cardinal F. S., *The Black Death of 1348 and 1349.* London: George Bell and Sons, 1908.

Getz, W. M. and J. Pickering, "Epidemic Models: Thresholds and Population Regulation," *American Naturalist*, 1983, 121:892–898.

Heide-Jørgensen, M. P. and T. Härkönen, "Epizootiology of the Seal Disease in the Eastern North Sea," *J. Appl. Ecol.*, 1992, 29:99–107.

Kermack, W. O. and A. G. McKendrick, "A Contribution to the Mathematical Theory of Epidemics," *Proc. R. Soc.*, 1927, A115:700–721.

Lenski, R. E. and R. M. May, "The Evolution of Virulence in Parasites and Pathogens: Reconciliation between Two Competing Hypotheses," *J. Theor. Biol.*, 1994, 169:253–265.

May, R. M., "Parasitic Infections as Regulators of Animal Populations," *American Scientist*, 1983, 71:36–45.

May, R. M. and R. M. Anderson, "Population Biology of Infectious Diseases: Part II," *Nature*, 1979, 280:455–461.

Osterhaus, A. D. M. E., J. Groen, P. DeVries, F. G. C. M. UytdeHaag, B. Klingenborn and R. Zarnke, "Canine Distemper Virus in Seals," *Nature*, 1988, 335:403–404.

Ross, R, "An Application of the Theory of Probabilities to the Study of *a priori* Pathometry, I, II, III," *Proc. R. Soc.*, 1916, A92:204–230, A93:212–225.

Swinton, J., J. Harwood, B. T. Grenfell and C. A. Gilligan, "Persistence Thresholds for Phocine Distemper Virus Infection in Harbour Seal *Phoca vitulina* Metappulations," *J. Anim. Ecol.*, 1998, 67:54–68.

Ziegler, Philip, *The Black Death*. London: Collins, 1969.

PROBLEMS AND EXERCISES

SI MODELS

1 (a) Default settings for the *Populus* Infectious Microparasitic Diseases Module run the *SI* model with density-dependent transmission. Parameter values are $b = 0.8$, $d = 0.7$, $\alpha = 0.9$, $\beta = 0.1$, $\upsilon = 0.9$, $S(0) = 20$, and $I(0) = 6$. First, set $I(0) = 0$ and run the model with the other parameters set to the defaults to see how this host population behaves in the absence of disease.

(b) Now return to the defaults with $I(0) = 6$; examine both *Populus* output plots and explain the dynamics of this example. Why do the densities behave as they do prior to settling at equilibria? If you need a hint, calculate the threshold susceptible host density required by this parasite.

(c) Do you find this case to be realistic? What are the minimum I densities that occur during the interval between times 3 and 7? What would happen if you had an initial susceptible density of $S(0) = 25$?

(d) Now set the initial susceptible density to $S(0) = 30$. Do you predict an initial oscillation or will this give a monotonic approach to the equilibrium? Discuss this question with your class, and explain your prediction. If you need a hint, determine the equilibrial prevalence (I/N) of the disease with these parameter values. Then run the new initial $I(0)$ and test your predictions.

(e) Parts b and d above hinted at "threshold" and "prevalence" hypotheses to explain the initial oscillation of this infectious-disease model. Can you think of another hypothesis and use additional simulation runs to test it? What are the biological attributes that contributed to the oscillation of single-species population-growth models?

2. (a) Recall the distinction between density-dependent and frequency-dependent transmission. Compare equation (6.1) with equation (6.23), and summarize the essential difference between them. Why do we call one density dependent and the other frequency dependent? What is the effect of frequency-dependent transmission on the threshold susceptible host density of the disease?

(b) Run the default case that you explored for problem 1 but shift to frequency-dependent transmission. What happens to the regulated equilibrium you observed previously with this case?

(c) Experiment with parameter values in an attempt to reestablish regulation. First, increase the disease prevalence, then reduce the host-population intrinsic growth rate. Are you able to regulate the host with frequency-dependent transmission? Use your own words to explain why.

(d) Is the interpretation of the disease-transmission parameter β exactly the same in host-microparasite models with density-dependent and frequency-dependent transmission? What are the units of β in each case?

SIR MODELS

3. Shifting to the SIR model with density-dependent transmission, use a pencil to calculate N_T, the threshold susceptible density required to sustain a case with $\alpha = 0.3, \beta = 0.5, \nu = 0.8, \gamma = 0, b = 0.25. d = 0.2, S(0) = 20, I(0) = 1,$ and $R(0) = 0$. Using this estimate of S_T, what will the initial R_0 of the disease be? What will R_0 be at equilibrium? Run a *Populus* simulation using these parameter values, and explain the screen output in your own words.

4. (a) What parameter changes could you make that would give a disease higher equilibrial prevalence? Look at the model schematic (Figure 6.7) and the basic equations (6.31)–(6.33) before you experiment with this. Are there simple adjustments that common sense suggests will lead to higher prevalence? Test your predictions, then initiate a simulation with the parameter values $\alpha = 0.9, \beta = 0.1, \nu = 0.9, b = 0.8, d = 0.7, S(0) = 20, I(0) = 6,$ and $R(0) = 6$. After the simulation equilibrates, estimate values from the screen for the susceptible, infected, and immune fractions, $I/N, I, N,$ and R/N. Then increase the transmission rate, β, and estimate any change in prevalence. Can you explain this result?

    (b) Notice that the default values for disease-induced mortality (α) and recovery (ν) are rather high. What would be the effect of lowering them? Increase the duration of infectiousness by lowering the recovery rate, and make new estimates of $I/N, I, N,$ and R/N. Was there any effect on the equilibrial prevalence? Now lower the disease-induced mortality (α) and make the same estimates of $I/N, I, N,$ and R/N. Can you explain these results?

    (c) Now increase the birth rate to $b = 0.9$. What happens to prevalence? Explain these results in your own words.

    (d) Can you think of any way to raise prevalence above its value in the previous example?

5. (a) Now that you have thought about prevalence, shift to the issue of host-population regulation. What are the factors that determine whether the parasite can hold its host to a stable population density? Run an SIR model with the default parameter values $[b = 0.8, d = 0.7, \alpha = 0.9, \beta = 0.1, \nu = 0.9, \gamma = 0, S(0) = 20, I(0) = 6,$ and $R(0) = 6]$ to observe an example of regulation. To see the converse, adjust natural mortality to a lower level, $d = 0.4$, and view the results.

    (b) Experiment with birth and natural death rate values (b and d) on one hand, and disease-induced mortality and prevalence (α and y) on the other, to get a feeling for the balance of forces that permit host regulation.

    (c) Now look back at equation (6.8) and the surrounding discussion. The parenthetic term in this equation illustrates the essential criterion for regulation. Use this criterion to select several different sets of parameter values, predict *a priori* whether their dynamics will be regulated, and check your predictions with simulation runs.

6. Running an *SIR* model with default values for all parameters except the disease-induced mortality rate, α, investigate the effect of virulence on a parasite's prevalence and ability to regulate host population density by filling in the following table:

α	Density-Dependent Transmission		Frequency-Dependent Transmission	
	N	y	N	y
0				
0.2				
0.4				
0.6				
0.8				
1.0				

A PROBLEM WITH EMPIRICAL DATA

7. Papers in the chapter references by Osterhaus *et al.* (1988), Heide-Jørgensen and Härkönen (1992), and Swinton *et al.* (1998) describe a 1988 outbreak of viral pneumonia among harbor seals, *Phoca vitulina*, in the North Sea adjacent to Denmark, Norway, and Sweden. After finding and reading these papers, use their data to parameterize a host-microparasite model. What parameter values have they measured directly? Are there other parameters, that can be estimated from their data? What are the units of each parameter? Do the data suggest that this is an interaction that can persist in equilibrium or is the long-term prospect more likely to be either the extinction of disease, or disease and host?

Glossary

Age structure. Proportional composition of a population with respect to the age of its members.

Amplitude. The difference between maximum and minimum sizes or densities in an oscillating population with regular expansions and contractions.

Asymptotic. An asymptotic function approaches some final value with time, such that the difference between current and final values declines by a constant fraction at each time step. The final value is the limit of the asymptotic function as time approaches infinity.

Birth rate. The average number of offspring produced per unit time, usually expressed on an individual basis as a *per capita* rate, b.

Carrying capacity. The population size, or density (size per unit area) that can be sustained in equilibrium with equal birth (b) and death (d) rates, usually parameterized as K.

Characteristic return time. A standard measure of the rate at which a regulated population returns to equilibrium after perturbation to higher or lower density, with a numerical value equal to the reciprocal of the intrinsic growth rate, $1/r$.

Cohort. The set of population members who are all approximately the same age.

Cohort-generation time. The average interval between the birth of a female and her offspring, calculated as the average of her ages when those offspring are born.

Competition coefficient. A measure of the strength of competitive interaction between species, quantified as the *per capita* effect of one species on the competitors' equilibrial density.

Competitive exclusion principle. The hypothesis, suggested by both theoretical models and empirical experiments, that two species cannot coexist if their resource requirements are identical.

Continuous. Referring to processes that are projected in infinitely small time increments; antonym of "discrete."

Death rate. The probability of dying per unit time, usually expressed on an individual basis as a *per capita* rate, d.

Definite integral. The sum of successive changes in a mathematical function evaluated over a finite interval.

Demography. The study of population age structure.

Density. Population size per unit area of habitat.

Density-dependent. Referring to population processes whose *per capita* rates increase or decrease with population density. For example, competition for limiting resources often increases the death rate and decreases the birth rate as population density increases.

Density-dependent transmission. An idealized characterization of disease transmission assuming that the number of new infections acquired from each infected host is directly proportional to the population density of susceptible hosts. It implies that the population is homogeneously mixed and that individuals contact each other randomly.

Deterministic. Refers to a model that produces a precise outcome, without incorporating effects of chance that would cause variation among successive simulations.

Differential equation. An equation describing variation in one parameter with respect to instantaneous changes in another, exemplified by the change in population size with respect to time, dN/dt.

Discrete. Referring to processes that are projected in finite time increments; an antonym of "continuous."

Discrete growth rate. The change in population size over a finite time interval, often expressed on a *per capita* basis.

Emigration. Departure of individuals reducing population size.

Equilibrium. A point where opposing processes balance. For example, when *per capita* birth and death rates are equal, the size of a closed population does not change.

Feedback. The effect of a model parameter value on the rate processes that cause changes in that parameter. For example, decreasing birth rates and increasing death rates may cause a negative feedback, slowing population growth as population size increases.

Finite-difference equation. An equation defining changes in one variable with respect to finite increments in another, exemplified by change in population size over a discrete interval.

Finite rate. A rate is the measure of one process occurring per unit of another, for example, the number of births per unit time. With a finite rate, the second process is measured in discrete, finite steps, as in births per year.

Function. A mathematical statement of relationship between two variables, such that specification of the value for one variable allows determination of the other.

Functional response. A predator's *per capita* rate of prey consumption as a function of prey population density. The functional response has an asymptotic upper limit if predators become satiated.

Global. Comprehensive. In theoretical ecology, a globally stable equilibrium is one to which a system will return from any set of nonzero population densities.

Immigration. Arrival of individuals increasing population size.

Instantaneous rate. A rate is the measure of one process occurring per unit of another, for example, the number of births per unit time. With an instantaneous rate, the second process is measured in infinitesimally small steps. The numerical value of an instantaneous rate must be defined over some finite interval. An instantaneous rate is the natural log of a finite rate defined over the same finite time interval.

Integrate. The process of adding successive changes in a mathematical function.

Interspecific. Between species.

Intraspecific. Within species.

Intrinsic rate of increase. The difference between *per capita* birth and death rates; the exponential growth rate of a population, usually parameterized as r.

Isocline. A line on the plot with densities of two interacting species on the x- and y-axes representing combined densities that eliminate net population growth for one species. Dynamics and equilibria of an interaction can be inferred when both species' isoclines are plotted together.

Iteroparous. A life history in which reproduction occurs more than once, often at regular intervals.

Life table. A list of age-specific survival and fertility expectations summarizing a populations' life history.

Limit cycle. An oscillation in the densities of two interacting species that returns to the same period and amplitude if it is perturbed to densities above or below the limit-cycle trajectory.

Logistic. A population growth model in which the *per capita* change of population size falls linearly with population density, producing a regulated equilibrium at the carrying capacity, $N = K$.

Mass action. The assumption that individuals in a population move or interact randomly so that properties of the whole population may be inferred from its density, without reference to individual characteristics.

Natural logarithm. A logarithm to the base $e \approx 2.718$.

Negative feedback. A negative effect of some model parameter value on the rate processes that cause changes in that parameter. The regulation of a logistic growth model is caused by the negative feedback of population size on population growth.

Net reproductive rate. In demography, the average number of female offspring produced by a female during her entire lifetime. In a model of microparasite dynamics, the average number of new infections caused by a single infected individual in a susceptible host population.

Neutral stability. The equilibrium of a Lotka-Volterra predator-prey model, with no tendency toward either regulation increasing oscillation. Neutrally stable interactions oscillate with a constant period and amplitude set by initial densities.

Oscillation. A regular cycle of changing population densities, with a period and amplitude.

Parameter. A variable or specified constant in a mathematical expression that determines the value or result.

Parthenogenetic. Asexual. In parthenogenetic species, females are able to produce progeny without mating. Several different cytogenetic mechanisms make this possible.

Per capita. Per individual, or literally, per head. A population-growth rate divided by population size gives population growth *per capita*.

Period. The time required for an oscillating population to retrace the same sequence of densities.

Population. A group of individuals of the same species in a defined area. The spatial extent of a population may be defined arbitrarily for the convenience of the investigator or it may have intrinsic natural limits.

Power series. A sum of successively higher powers of some parameter each multiplied by a constant, as in a Taylor Series.

Prevalence. The proportion of host population members infected by a disease.

Quadratic. A second-degree equation, for example, $ax^2 + bx + c = 0$, where a, b, and c are constants.

Random. Haphazard, having no recognizable pattern. A phenomenon with an *a priori* expectation between 0 and 1. In a random sample of individuals from a population, each member has an equal probability of being drawn.

Regulate. To confer stability on a dynamic equilibrium, so a system returns to that equilibrium after a perturbation.

Semelparous. A life history in which organisms reproduce only once.

Semi-log plot. A graph with additive increments on the x-axis and multiplicative increments on the y-axis.

Sigmoid. S-shaped. A time trajectory of density-dependent growth (N vs. t) increases exponentially at low density when negative feedback is weak, then makes an asymptotic approach to the carrying capacity, producing an S-shaped plot.

Size. The number of individuals in a population, often quantified as a density; the number of individuals per unit area.

Stable. A term suggesting constancy. A stable equilibrium density is one to which a regulated population returns after being perturbed to higher or lower densities. A stable cycle returns to the same period and amplitude after similar perturbation.

Stable age distribution. A steady population age structure that arises when age-specific survival and fertility remain constant.

Stochastic. Effects of chance and random sampling. A stochastic simulation may produce different results in successive runs with the same parameter values, attributable to these effects of chance.

Time series. A set of sequential estimates of some population parameter, showing successive changes over time.

Zero-net-growth isocline. A line on a graph of population dynamics that represents sets of population densities or resource concentrations causing birth and death rates of one participating species to balance.

Index

A

Agave kaibabensis (century plants), 63
Age-structured populations, 52–55. *See also*
    Demography
  cohort-generation time, 53–54
  Cole's paradox and evolution of life his-
    tories, 55–57
  life-cycle graphs for, 62–65
  Lotka-Euler equation, 54–55, 58
  Populus simulations of, 46, 48
  projection matrices for, 58–61
  reproductive value, 57–58
  stage-structured populations vs., 63
AIDS, 115
Amplitude, 28
Ant-rodent competition, 73
Asterionella formosa, 71–73
Asymptotic approach to extinction, 7–9
Asymptotic decline in population, 22

B

Birth, 1
  per capita, 23
Black-headed budworm, 32, 33

C

Carrying capacity(ies)
  in continuous logistic model, 17–18
  predator-prey dynamics and, 102
  probability of coexistence and, 84–85
Century plants (*Agave kaibabensis*), 63
Chicken pox, 128

Cohort-generation time, 53–54
Cohorts, 41. *See also* Discrete logistic mod-
    els of density-dependent population
    growth
  geometric growth of, 4–5
Cole's paradox, 55–57
Competition
  defined, 71
  interspecific, 71
  intraspecific, 71
  Lotka-Volterra, 71–90
    coexistence vs. displacement outcomes
      of, 73, 79–84
    competition coefficient, 73
    dynamics of, 74–76
    empirical examples of, 71–73
    isocline analyses, 77–84
    simplifications in, 86
Continuous population growth, 2
Continuous population growth models
  for density-dependent populations, 16–25
    carrying capacity in, 17–18
    dynamics and equilibria in, 20–24
    per capita vs. whole population growth
      rate, 19
    population doubling time, 24–25
    Populus simulation of, 19–20, 21, 24
    range of values allowed in, 25
  discrete models compared with, 5
Continuous predator-prey models, 91–114
  empirical example, 91–92
  Lotka-Volterra, 92–104, 111
    assumptions of, 92–93, 102, 104
    density-dependent prey and, 101–4,
      111
    dynamics of, 94, 100–101

interspecific components of, 93–94
intraspecific components of, 93
isocline analyses in, 94–101
theta-logistic, 104–11
functional responses in, 105–9
interspecific component of, 104
intraspecific component of, 104, 110
paradox of enrichment, 109–10

D

Death, 1
per capita, 23
Demography, 41–52. *See also* Age-structured populations
constant survival and fertility schedules,
49–52
estimating survival and fertility parameters, 42–46
population projection from life table,
46–49
Density-dependent feedback, 12, 71
detecting, 33–35
Density-dependent population growth, 16–40
continuous logistic model of, 16–25
carrying capacity in, 17–18
dynamics and equilibria in, 20–24
per capita vs. whole population growth
rate, 19
population doubling time, 24–25
range of values allowed in, 25
simulating with *Populus*, 19–20, 21, 24
defined, 16
detecting density dependence, 33–35
discrete logistic models of, 29–32
critical values for, 31
dynamics and equilibria, 31–32
in *Populus*, 29–31
lagged logistic model of, 25–28
critical values for, 28
dynamics and equilibria in, 27–28
negative feedback in, 16, 17–19
Density-dependent prey, modifying Lotka-
Volterra model with, 101–4

Density-dependent transmission of disease,
116
SI model with, 116–22
dynamics and equilibria in, 117–22
SIR model with, 125–26
Density-independent population growth
model, 1–15, 34
assumptions of, 1–2
continuous vs. discrete models, 5
defined, 1
descriptive power of, 10–12
exponential growth with continuous
breeding, 2–3
geometric growth of discrete cohorts, 4–5
Populus simulation of, 6–10, 11
Diatoms, 72–73
Differential equation, 2
Discrete changes in population size, 4
Discrete-growth rate, 5
Discrete logistic models of density-dependent population growth, 29–32
continuous models compared with, 5
critical values for, 31
dynamics and equilibria, 31–32
in *Populus*, 29–31
Doubling time, 24–25
Dynamics
in continuous logistic model, 20–24
in discrete logistic models of density-
dependent population growth,
31–32
in lagged logistic model of density-dependent population growth, 27–28
in *SI* model of infectious microparasitic
disease
with density-dependent transmission,
117–22
with frequency-dependent transmission,
122–25

E

Emigration, 1
Enrichment, paradox of, 109–10

Equilibrium/equilibria
  in logistic models
    continuous, 20–24
    discrete, 31–32
    lagged, 27–28
  in Lotka-Volterra competition, 77–79
  of Lotka-Volterra Predator-Prey equations, 94–101
  monotonic approach to, 28
  in *SI* model of infectious microparasitic disease with density-dependent transmission, 117–22
Euler's (Lotka's) equation, 54–55, 58
Evolution
  of life histories, 55–57
    reproductive value, 57–58
  of virulence, 129–30
Exponential growth, 7
  with continuous breeding, 2–3
Extinction, asymptotic approach to, 7–9

F
Feedback
  density-dependent, 12, 71
    detecting, 33–35
  negative, 16, 17–19, 71. *See also* Carrying capacity(ies)
Fertility-parameter estimates, 42–46
  for age-structured population growth, 61
Fertility schedules, constant, 49–52
Finite difference equation, 4
Finite-growth rate, 5
Frequency-dependent transmission of disease, *SI* model with, 122–25
  dynamics of, 122–25
Functional response(s), 93–94
  in theta-Logistic Predator-Prey Model, 105–9
    Type I, 15, 108, 109
    Type II, 106–9
    Type III, 106
Functions, 2, 16
  quadratic, 19

G
Genes, age-specific effects, 58
Geometric growth, 7
  of discrete cohorts, 4–5
Globally stable equilibrium, 23

H
Hare-lynx predator-prey dynamics, 91–92. *See also* Lotka-Volterra predator-prey models
Host-parasite interactions. *See* Infectious microparasitic diseases
Human immunodeficiency virus (HIV), 115

I
Immigration, 1
Immunization, dynamics of, 126–29
  smallpox eradication, 128–29
Infectious microparasitic diseases, 115–33
  Anderson and May model of, 115–16
  evolution of virulence, 129–30
  immunization dynamics, 126–29
    smallpox eradication, 128–29
  SI model of, 116–25
    with density-dependent transmission, 116–22
    with frequency-dependent transmission, 122–25
  SIR model with density-dependent transmission, 125–26
Instantaneous growth rate (intrinsic rate of increase), 2
Instantaneous rates, 2, 3
Integration, 3
  numerical, 27
Interspecific competition, 71
Intraspecific competition, 71
Intrinsic growth rate, 24, 118
Intrinsic rate of increase (instantaneous growth rate), 2
Isocline analyses

of Lotka-Volterra competition, 77–84
coexistence vs. displacement outcomes in, 79–84
in Lotka-Volterra predator-prey models, 94–101
Iteroparous organisms, 54, 55–57

L

Lagged logistic model of density-dependent population growth, 25–28
critical values for, 28
dynamics and equilibria in, 27–28
Lepus americanus (snowshoe hare), 91–92
Life-cycle graphs for age-structured populations, 62–65
Life histories, evolution of, 55–57
reproductive value, 57–58
Life table, 43
population projection from, 46–49
Limit cycle, 108–9
stable, 28
Logarithm, natural, 3
Logistic models of density-dependent population growth, 16–28
continuous, 16–25
discrete, 29–32
lagged, 25–28
Lotka-Euler equation, 54–55, 58
Lotka-Volterra competition, 71–90
coexistence vs. displacement outcomes of, 73, 79–84
carrying capacities and, 84–85
competition coefficient, 73
dynamics of, 74–76
empirical examples of, 71–73
isocline analysis, 77–84
simplifications in, 86
Lotka-Volterra predator-prey models, 92–104, 111
assumptions of, 92–93, 102, 104
density-dependent prey and, 101–4, 111
dynamics of, 94, 100–101
interspecific components of, 93–94
intraspecific components of, 93

isocline analyses in, 94–101
Lynx-hare predator-prey dynamics, 91–92. *See also* Lotka-Volterra predator-prey models

M

Macroparasites, 115
Mass-action infection, 116
Matrix projections, 58–61
for stage-structured populations, 62–65
Matrix × vector multiplication, 59–60
Microparasites, 115. *See also* Infectious microparasitic diseases
Monotonic approach to equilibrium, 28
Myxoma virus, 115, 129

N

Natural logarithm, 3
Negative feedback, 16, 17–19, 71. *See also* Carrying capacity(ies)
Net reproductive rate of disease, 52, 117–18, 127
virulence and, 129
Neutrally stable dynamics, 100–101
Numerical integration, 27
Numerical response, 94

O

Odocoileus virginianus. See White-tailed deer
Oryctolagus cuniculus, 115
Oscillations, 28
density-dependent predator-prey, 102-3
neutrally stable, 101
Ovulation rates of white-tailed deer, 16, 17

P

Paradox of enrichment, 109–10

Parameters, 2
Parasites. *See* Infectious microparasitic diseases
Parthenogenetic females, 1
Per capita growth, 3
    in continuous logistic model, 23
    declining, 18–19
    whole population growth rate vs., 19
Perennial plants, 62
Period, 28
Phase plane, 76
Poliomyelitis, 127–28
Population density, 1
Populations, 1
    defined, 1
    regulated, 23–24
Population size, 1
    discrete changes in, 4
Populus simulation(s)
    of age-structured population growth, 46, 48
    of continuous logistic model, 19–20, 21, 24
    of density-dependent population growth, 34
    of density-independent population growth, 6–10, 11, 34
    of discrete logistic model, 29–31
    illustrating paradox of enrichment, 110
    of lagged logistic growth, 26–27
    of Lotka-Volterra competition, 74–76
        isocline graphs, 84, 85
    of Lotka-Volterra predator-prey model, 95–96, 100
        with density-dependent feedback, 102–3
    of microparasite disease, 115–16
    of resource competition, 86
    of *SI* host-microparasite model with density-dependent transmission, 121–22
    of theta-Logistic predator-prey model, 106, 107, 108–9
Power series, 39
Predator-prey models, continuous, 91–114
    empirical example, 91–92
    Lotka-Volterra, 92–104, 111
        assumptions of, 92–93, 102, 104
        density-dependent prey and, 101–4, 111
        dynamics of, 94, 100–101
        interspecific components of, 93–94
        intraspecific components of, 93
        isocline analyses in, 94–101
    theta-logistic, 104–11
        functional responses in, 105–9
        interspecific component of, 104
        intraspecific component of, 104, 110
        paradox of enrichment, 109–10
Prevalence of disease, 118
Projection matrices
    for age-structured populations, 58–61
    for stage-structured populations, 62–65

Q
Quadratic function, 19

R
Regulated population, 23–24
Reproductive value, 57–58
Rodent-ant competition, 73
Rubella, 127

S
Seasonally reproducing species, density-independent growth model for, 4–5
Semelparous organisms, 54, 55–57
Sigmoid (S-shaped) time trajectory, 22
SI model of infectious microparasitic disease, 116–25
    with density-dependent transmission, 116–22
    dynamics and equilibria in, 117–22
    with frequency-dependent transmission, 122–25
    dynamics of, 122–25
SIR model with density-dependent transmission, 125–26

Smallpox, eradication of, 128–29
Stable limit cycle, 28
Stage-structured populations, 62–65
    age-structured populations vs., 63
Survival-parameter estimates, 42–46
    for age-structured population growth,
        60–61
Survival schedules, constant, 49–52
Susceptible host density threshold, 118
Sylvilagus brasiliensis, 115
Synedra ulna, 71–73

T
Territorial animals, feedback function for,
    19
theta-Logistic Predator-Prey Model,
    104–11
    functional responses in, 105–9
    interspecific component of, 104
    intraspecific component of, 104, 110
    paradox of enrichment, 109–10
Time series of density estimates, 33
Transmission of disease
    density-dependent, 116–22
    frequency-dependent, 122–25

V
Vaccination, 128
Variolation, 128
Variola virus, 128–29
Vectors, 59
Virulence, evolution of, 129–30

W
White-tailed deer, 4
    age-specific reproductive value for, 57
    age structure of, 43
    constant survival and fertility schedules
        for, 49–52
    estimating survival and fertility parame-
        ters for, 42–46
    ovulation rates of, 16, 17
    population projection from life table for,
        46–49
    schematic of life history for, 42, 49
    seasonal births and deaths of, 41–42

Y
Yeast cells, 71